SciQuest

Addy's Adventure

SciQuest: Addy's Adventure

First Edition 2026

Email: support@elementalscience.com

ISBN #978-1-953490-38-4

Printed In USA For worldwide distribution

For more copies write to :
Elemental Science
PO Box 4583
Maryville, TN 37802
support@elementalscience.com

Copyright Policy

Quick Start Guide

It's time to head out on a quest—a scientific quest—a SciQuest.

What Will We Learn?

This semester, you will follow Addy and her family as they travel down Route 66, learning about matter, reactions, forces and and more. See p. 13 for a list of the topics explored in this program.

What Do I Need?

In addition to this activity guide, you will need the following materials:

1. **Novel –** *The Route 66 Sticky Note Quest* - Each week, a chapter in this book will spark the student's interest to learn more. You can get the paperback novel, the Kindle version, or the audiobook.
2. **Age-appropriate Encyclopedias –** The bulk of the scientific information will come from these. We recommend that you have at least one or more of these on hand. If you can only purchase one or two of these, we recommend getting the ones with an *.
    - *Usborne Science Encyclopedia**
    - *Kingfisher Science Encyclopedia**
    - *Usborne Illustrated Dictionary of Science*
    - *Everything You Need to Ace Science in One Big Fat Notebook*
3. **Student Materials –** You will need to purchase *SciQuest: Level 200 - Addy's Adventure.* See inside the lab journal on p. 8 of this guide.
4. **Demonstration Supplies –** See p. 14 for a full list, or save yourself time and get the *SciQuest Year 1 Experiment Kit*, which includes the materials for both level 100 and 200.

What Will a Week Look Like?

Each week, you and your middle school student will:

- ✓ **Read** a chapter in a middle-grade fiction book with science hints and pick one or more of the assigned pages, discussing the information with the included questions.
- ✓ **Do** the weekly experiments by following the procedure, recording your observations and results, and drawing conclusions about what you saw.
- ✓ **Write** down what the students have learned in a way that is appropriate for their skills and label a related sketch.

You can also add in the memory work and additional science activities if you want to dig deeper into a topic. For a more detailed explanation of the components in each lesson, we highly recommend checking out the peek inside this program on pp. 6-7 and reading the introduction starting on pp. 9-12. Otherwise, the first lesson begins on p. 15.

Table of Contents

TEACHER NOTES AT-A-GLANCE

The teacher guide is your go-to resource for guiding your students through science!

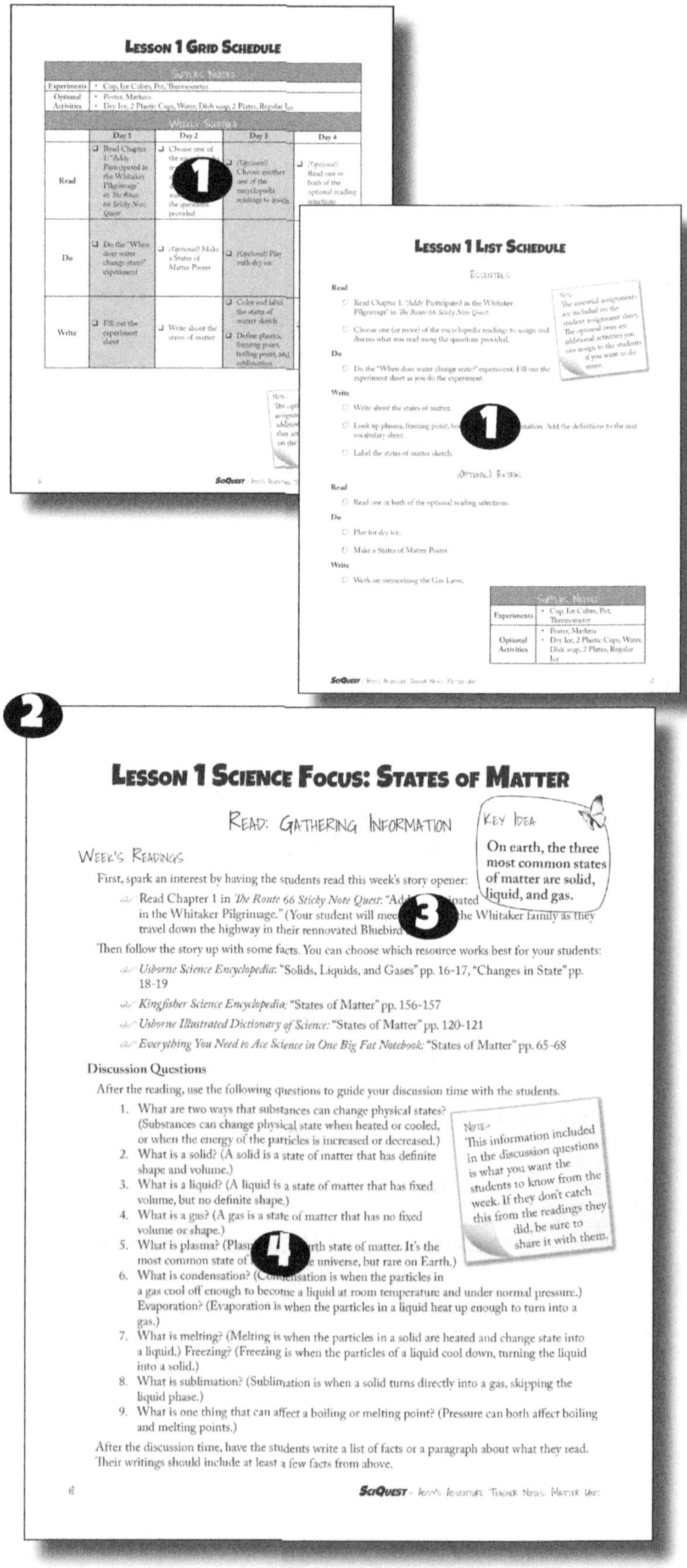

1. TWO SCHEDULING OPTIONS

Provide guidance to the students with these two grid-style scheduling options. There are two-day-a-week and five-day-a-week schedules. These schedules break down the essential work and the optional activities into manageable chunks so that you can help them proceed with confidence.

2. NOTES

Get key information and further explanations of the topics when necessary after the schedules. These notes will assist you as you teach science to your students and are described below.

READ SCIENCE

3. READING ASSIGNMENTS

See the week's reading assignments. First, you will find the chapter from the springboard novel. After that, you will find the encyclopedia pages, which contain the core of the information for the week.

4. DISCUSSION QUESTIONS WITH ANSWERS

Get discussion questions with the answers so you have the tools necessary to lead an effective discussion time without having to read the selection.

5. OPTIONAL READINGS

Find optional books you can check out from your local library.

Teacher Notes At-A-Glance

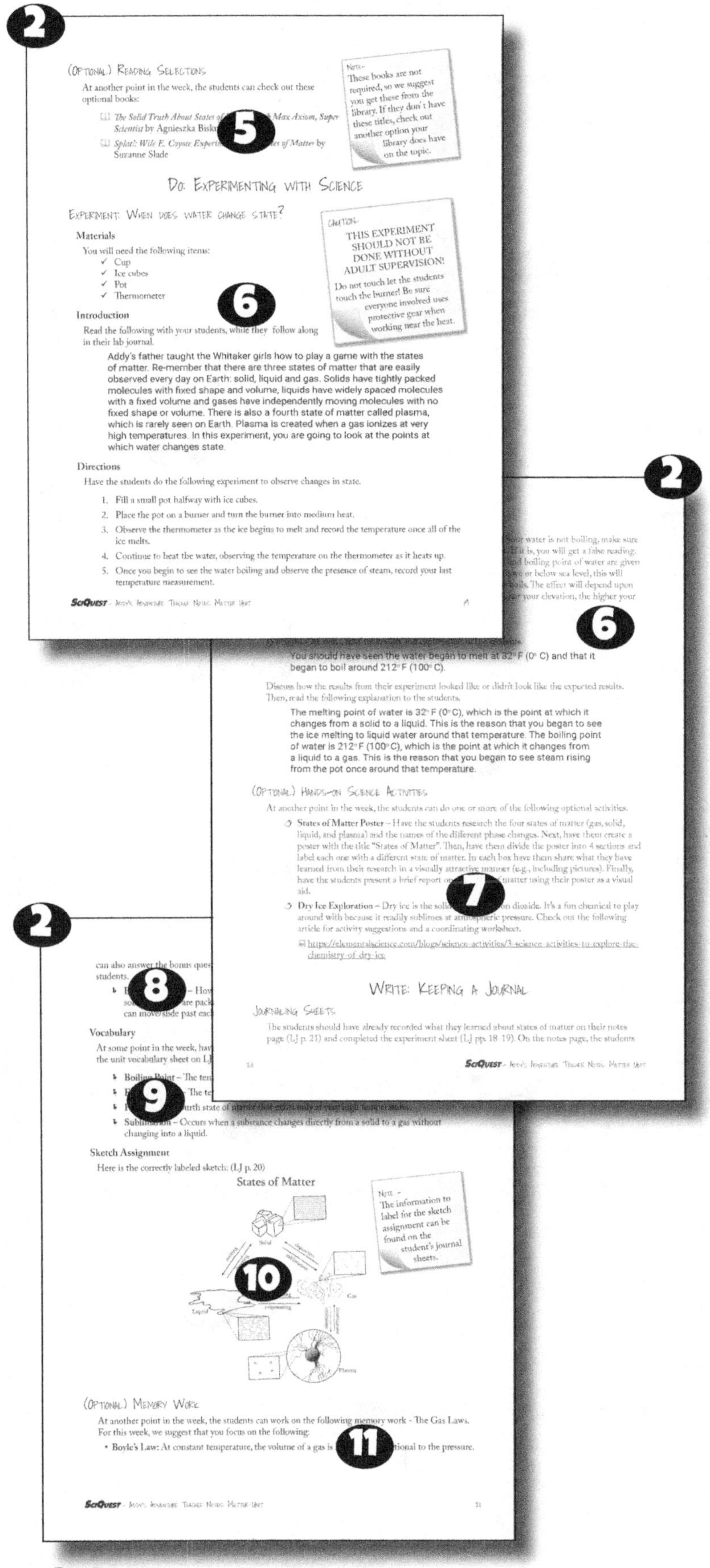

Do Science

6. Experiment Information

Know what you need to know about the weekly experiment, which will coordinate with the topic. This section includes the introduction and procedure, which are also included in the student materials. It also lists the expected results, an explanation, and ideas to take the hands-on science fun further.

7. Optional Hands-on Activities

Add a bit of fun with these optional research projects and hands-on science project ideas. These relate to and reinforce what the students are learning about the week's subject.

Write Science

8. Journal Sheets

Understand what the students will be writing down for the week, along with a bonus question. These are deeper inquiries designed to make the students think about the subject matter.

9. Vocabulary

Enhance your students' science vocabulary by introducing words that are relevant to the weekly topic.

10. Sketch Answers

Know how and where the students should label the pre-drawn sketches with the completed images.

11. Optional Memory Work

Have the students memorize key facts about the unit's topics with these optional memory work selections.

Lab Journal At-A-Glance

Go on a scientific quest to learn about matter, solutions, and forces.

1. Unit Memory Work

   Memorize key facts about the unit's topics with these optional memory work selections.

2. Unit Glossary

   Keep all the unit vocabulary in one place. The glossary gives space to write the definitions for each vocabulary word.

3. Student Assignment Sheet

   See the plan for the week on the student assignment sheets, broken into reading, doing, and writing sections. This sheet has all the tasks for the week with checkboxes for you to use as you complete the work.

4. Experiment Sheet

   Do an experiment related to the topic for the week. The Experiment Sheet contains the introduction and the procedure for the experiment. It also has space to record the hypothesis, materials, observations, results, and conclusion.

5. Pre-drawn Sketch

   Color the pre-drawn sketch relating to the week's topic. Then, label it with the information given on the page.

6. Written Assignment

   Know the questions for discussion time, and have plenty of space to write either a list of facts, an outline, or a summary.

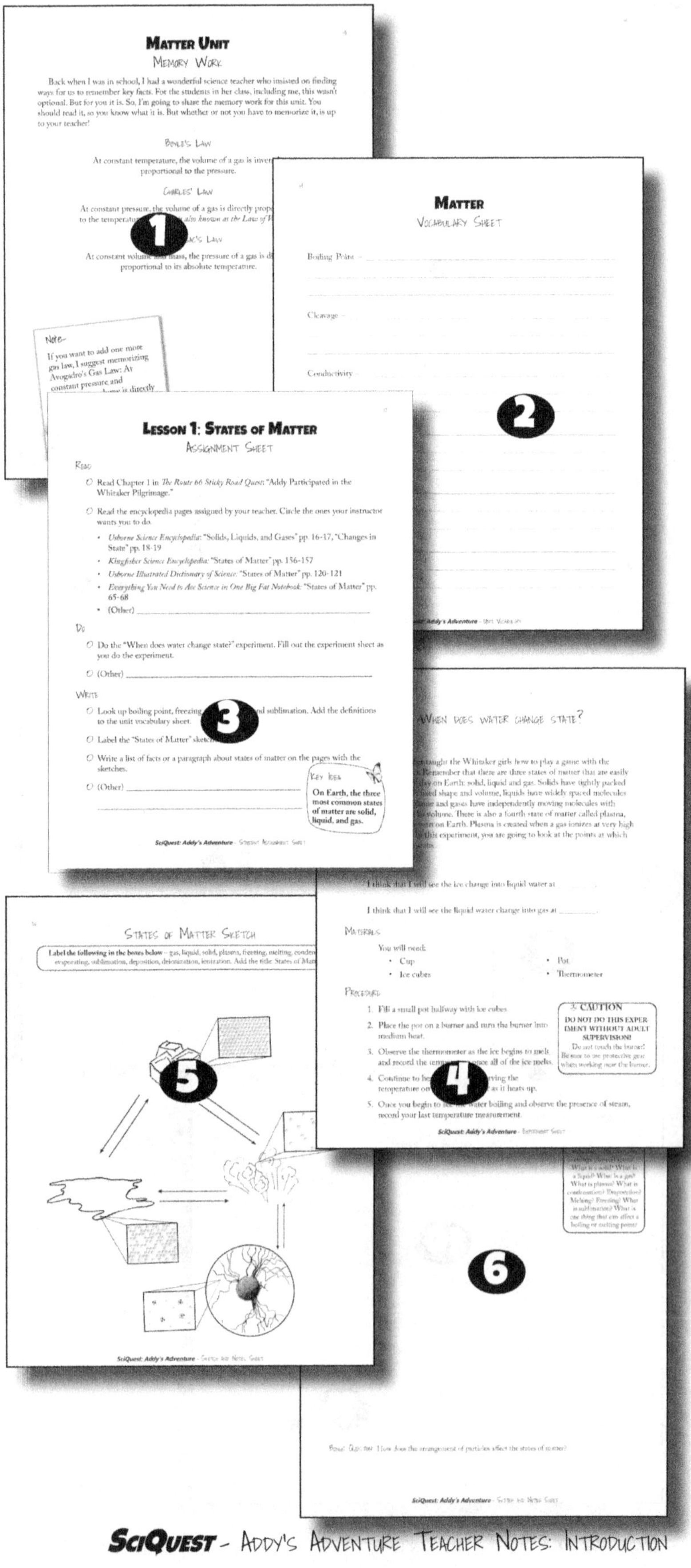

SciQuest: Addy's Adventure Introduction

Welcome to a quest to learn about science, or should I say a SciQuest! This is a journey that your students are going to embark on with a bit of independence.

In *SciQuest 200: Addy's Adventure*, you'll be learning about matter, solutions, reactions, forces, and energy. Addison (Addy) Whitaker will serve as an introduction for your students for the week. They will learn about the topics in chemistry and physics by reading, doing, and writing about science.

The Broad-Brush-Stroke View

In a nutshell, your students will enjoy learning about science as they do three things each week. These three things are based on the *Three Keys to Teaching Science* by Paige Hudson. Check them out here: https://elementalscience.com/blogs/news/3-keys. Here is what they are:

Read: Gathering Information

In this section of the lesson notes, you will find the assigned readings and discussion questions. Check out this podcast to understand the importance of discussion time:

- Don't skip that discussion time - https://elementalscience.com/blogs/podcast/53

You'll also find a few optional assignments for those students who want to dig deeper.

Do: Experimenting with Science

In this section of the lesson notes, we will be employing those inquiry-based methods. In other words, the students are going to do science. You will find an experiment introduction and directions – don't worry, these are in the student's journal, too. You will also find the expected results, an explanation, and any troubleshooting tips for the experiment in this section. Read this article to understand how your role is shifting from presenter-of-hands-on-science to mentor-of-a-budding-scientist:

- Scientific demonstrations vs. experiments - https://elementalscience.com/blogs/podcast/94

Plus, we added a few optional activities for those students who love to play with science.

Write: Keeping a Journal

In this section of the lesson notes, we are helping our students to organize what they have learned and to think analytically about it. This is where you will find the written assignments, vocabulary, and memory work for the lesson. The written assignments include a customizable writing assignment (either a list of facts, an outline, or a paragraph) and the sketch for the lesson. Watch this series of videos to learn about the progression of writing skills as they pertain to science:

- Writing in Homeschool Science: The Elementary Years - https://youtu.be/BrunFyeHhlQ
- Writing in Homeschool Science: Middle School and Beyond - https://youtu.be/B96DKXriCng

Also, with the written work, we have included a bonus question to help the students start to make those all-important connections.

The Nitty-Gritty Details of the Lesson Notes

Week's Readings

First, spark an interest by having the students read this week's story opener. These chapters come from the middle-grade novel, *The Route 66 Sticky Note Quest*. Your students will read about Addy and her family as they travel down America's Mother Road, Route 66. Each chapter has a hint of science and is designed to be read by the students on their own.

Later in the week, the students will follow up the story with facts. The following encyclopedias have pages detailed in this guide. You can choose which resources work best for your student. They will be used in subsequent levels of this program. If you can only purchase one or two of these, we recommend getting the ones with an *.

- *Usborne Science Encyclopedia**
- *Kingfisher Science Encyclopedia**
- *Usborne Illustrated Dictionary of Science*
- *Everything You Need to Ace Science in One Big Fat Notebook*

Discussion Questions

After all that reading, we have provided questions, along with the answers, to guide your discussion time with the students.

(Optional) Reading Selections

In this section, you will find additional reading suggestions. These are not required, but these library books can be used to enhance what the students have learned.

Experiment Information

Materials

This section will detail the supplies you need for the experiment. As much as possible, we have attempted to make these household supplies. However, if you don't want to pull these items together, you can purchase an experiment kit from us that contains the supplies for level 100 and 200.

Introduction

This section shares a glimpse of the science behind the experiment. It is also included in the lab journal.

The paragraphs written in this font are meant to be read to the student.

Directions

This section will detail what to do for the experiment, step by step. It is also included in the lab journal.

Results and Explanation

When the students are done with the experiment, read the results and explanation to them. Discuss how the results from their experiment looked like or didn't look like the expected results.

As before, the paragraphs written in this font are meant to be read to the student.

(Optional) Hands-on Science Activities

In this section, you will find optional hands-on science activities to reinforce what the students are learning.

Write: Keeping a Journal

Journaling Sheets

This section will have the page numbers and bonus questions. The students should record what they learn about the subjects on notes pages and the experiment sheets in the lab journal. On the notes page, the students can also answer the bonus question. These questions are meant to challenge the students and cause them to think about the subject.

Vocabulary

The students will look up the vocabulary words and add the definitions to their unit vocabulary sheet in the lab journal. In this guide, you will find the definitions for the words they have to look up.

Sketch

In this section, you will find the labeled sketches to use as an answer key.

(Optional) Memory Work

Every unit has memory work, which you can choose to assign or not.

Additional Resources

We have put together a page containing quick links to the activities suggested in this guide along with several helpful downloads:

💻 https://elementalscience.com/blogs/resources/sq200

How to Schedule this Study

SciQuest 200: Addy's Adventure should take up to three hours per week to complete. You and your student can choose whether to complete the work over five days or over two days. Below are two options for scheduling to give you an idea of how you can schedule your week:

Two-days-a-week Schedule

Day 1

Define the vocabulary, do the experiment, and complete the experiment sheet.

Day 2

Read the assigned pages and discuss them together, take your notes, and complete the sketch.

Five-days-a-week Schedule

Day 1

Do the experiment and complete the experiment sheet.

Day 2

Record the dates and define the vocabulary.

Day 3

Read the assigned pages, discuss them, and complete the sketch.

Day 4

Prepare the science report or an outline.

Day 5

Complete one of the other activities assigned by your teacher and work on the memory work.

Final Thoughts

If you find that this program contains too much work, please tailor it to the needs of your students. With that said, I encourage you to contact us with any questions or problems that you might have concerning *SciQuest 200: Addy's Adventure* at support@elementalscience.com. Our team will be more than happy to answer them as soon as we are able. I hope that you and your students enjoy your quest to learn about science!

~ Paige Hudson

List of Topics

Matter Unit

- States of Matter
- Solid Structures
- Crystals
- Density
- Surface Tension
- Intermolecular Forces
- Gas Laws
- Diffusion
- Kinetic Theory

Solutions and Reactions Unit

- Molecules
- Compounds
- Mixtures
- Polarity
- Solutions
- Filtration
- Chromatography
- Electrolysis
- Chemical Bonding
- Chemical Reactions

Scientist Biography Unit

- The students will learn about a scientist of their choice.

Forces and Motion Unit

- Forces
- Kinetic Energy
- Potential Energy
- Work
- Air Resistance
- Friction
- Gravity
- Fluid Dynamics
- Pressure
- Inertia
- Mass
- Weight
- Motion
- Speed
- Acceleration
- Velocity
- Simple Machines

Demonstration Supplies

Matter Unit

Lesson	Materials Needed
1	Cup, Ice Cubes, Pot, Thermometer
2	Wide mouthed jar, String, Pencil, Pipe cleaner, Water, Borax, Scissors
3	4 Clear cups, Eye dropper, Table salt, Food coloring, Water
4	2 Cups, Apple juice, Timer, Partner

Reactions Unit

Lesson	Materials Needed
5	Bag of multi-colored marshmallows, Toothpicks
6	5 Clear cups (or beakers), 5 Plastic spoons, Sugar, Salt, Baking soda, Flour, Petroleum jelly, Water, Vegetable oil, Tablespoon
7	4 Coffee filters, Markers, Alcohol, Coffee can or a wide-mouthed jar, Rubber bands, Eye dropper
8	Distilled water, 2 Test tubes, Salt, Glass cup, 2 Alligator clips, Covered copper wire, 6v Lantern battery, Permanent marker
9	Cake frosting, Red and yellow bite-sized candies (such as regular sized M&M's)
10	Yeast, Hydrogen peroxide, Epsom salts, Water, 2 Cups, 2 Thermometers

Forces Unit

Lesson	Materials Needed
12	Thick, sturdy cardboard, 1 Brad fastener, Rubber band, Hole punch or nail, String – about 3 in (10 cm), 3 Jumbo paper clips, Pen, Objects of varying weight
13	Goldfish cracker, Small marshmallow, Piece of lettuce, Piece of bacon fat, Aluminum pan, Matches, Safety glasses, Bucket of water
14	Force Meter from Lesson 12, Small wooden block (aka. Jenga block), Eye-hook screw, Sandpaper, Felt, Foil, Spray oil, Tape measure
15	2 Tin cans (one large, one small), Screw, Water, Piece of clay, Cup measurer, Tape measurer
16	Jenga block with the eyehook from last week, String, 2 Toy cars, Egg
17	Cardboard or plastic track, Blocks or thick books, Toy car, Stopwatch
18	Materials will vary based on what you choose to build

SciQuest

Addy's Adventure Teacher Notes

Matter Unit

What you'll study in this unit-

Memory Work-

Three Gas Laws

(LJ* p. 13)

*LJ = Lab Journal

Lesson 1 Grid Schedule

Supplies Needed	
Experiments	• Cup, Ice Cubes, Pot, Thermometer
Optional Activities	• Poster, Markers • Dry Ice, 2 Plastic Cups, Water, Dish Soap, 2 Plates, Regular Ice

Weekly Schedule

	Day 1	Day 2	Day 3	Day 4
Read	❑ Read Chapter 1: "Addy Participated in the Whitaker Pilgrimage" in *The Route 66 Sticky Note Quest*	❑ Choose one of the encyclopedia readings to assign and discuss what was read using the questions provided	❑ *(Optional)* Choose another one of the encyclopedia readings to assign	❑ *(Optional)* Read one or both of the optional reading selections
Do	❑ Do the "When does water change state?" experiment	❑ *(Optional)* Make a States of Matter poster	❑ *(Optional)* Play with dry ice	
Write	❑ Fill out the experiment sheet	❑ Write about the states of matter	❑ Color and label the states of matter sketch ❑ Define plasma, freezing point, boiling point, and sublimation	❑ *(Optional)* Work on memorizing the Gas Laws

Note–
The optional assignments are additional activities that are not included on the student's assignment sheet.

Lesson 1 List Schedule

Essentials

Read

- ○ Read Chapter 1: "Addy Participated in the Whitaker Pilgrimage" in *The Route 66 Sticky Note Quest*.
- ○ Choose one (or more) of the encyclopedia readings to assign and discuss what was read using the questions provided.

Note-
The essential assignments are included on the student assignment sheet. The optional ones are additional activities you can assign to the students if you want to do more.

Do

- ○ Do the "When does water change state?" experiment. Fill out the experiment sheet as you do the experiment.

Write

- ○ Write about the states of matter.
- ○ Look up plasma, freezing point, boiling point, and sublimation. Add the definitions to the unit vocabulary sheet.
- ○ Label the states of matter sketch.

(Optional) Extras

Read

- ○ Read one or both of the optional reading selections.

Do

- ○ Play with dry ice.
- ○ Make a States of Matter poster.

Write

- ○ Work on memorizing the Gas Laws.

Supplies Needed	
Experiments	• Cup, Ice Cubes, Pot, Thermometer
Optional Activities	• Poster, Markers • Dry Ice, 2 Plastic Cups, Water, Dish Soap, 2 Plates, Regular Ice

Lesson 1 Science Focus: States of Matter

Read: Gathering Information

Key Idea

On earth, the three most common states of matter are solid, liquid, and gas.

Week's Readings

First, spark an interest by having the students read this week's story opener:

- Read Chapter 1 in *The Route 66 Sticky Note Quest*: "Addy Participated in the Whitaker Pilgrimage." (Your student will meet Addy and the Whitaker family as they travel down the highway in their rennovated Bluebird bus.)

Then follow the story up with some facts. You can choose which resource works best for your students:

- *Usborne Science Encyclopedia*: "Solids, Liquids, and Gases" pp. 16-17, "Changes in State" pp. 18-19
- *Kingfisher Science Encyclopedia:* "States of Matter" pp. 156-157
- *Usborne Illustrated Dictionary of Science:* "States of Matter" pp. 120-121
- *Everything You Need to Ace Science in One Big Fat Notebook:* "States of Matter" pp. 65-68

Discussion Questions

After the reading, use the following questions to guide your discussion time with the students.

1. What are two ways that substances can change physical states? (Substances can change physical state when heated or cooled, or when the energy of the particles is increased or decreased.)
2. What is a solid? (A solid is a state of matter that has definite shape and volume.)
3. What is a liquid? (A liquid is a state of matter that has fixed volume, but no definite shape.)
4. What is a gas? (A gas is a state of matter that has no fixed volume or shape.)
5. What is plasma? (Plasma is the fourth state of matter. It's the most common state of matter in the universe, but rare on Earth.)
6. What is condensation? (Condensation is when the particles in a gas cool off enough to become a liquid at room temperature and under normal pressure.) Evaporation? (Evaporation is when the particles in a liquid heat up enough to turn into a gas.)
7. What is melting? (Melting is when the particles in a solid are heated and change state into a liquid.) Freezing? (Freezing is when the particles of a liquid cool down, turning the liquid into a solid.)
8. What is sublimation? (Sublimation is when a solid turns directly into a gas, skipping the liquid phase.)
9. What is one thing that can affect a boiling or melting point? (Pressure can both affect boiling and melting points.)

Note–
This information included in the discussion questions is what you want the students to know from the week. If they don't catch this from the readings they did, be sure to share it with them.

After the discussion time, have the students write a list of facts or a paragraph about what they read. Their writings should include at least a few facts from above.

(Optional) Reading Selections

At another point in the week, the students can check out these optional books:

- *The Solid Truth About States of Matter With Max Axiom, Super Scientist* by Agnieszka Biskup
- *Splat!: Wile E. Coyote Experiments with States of Matter* by Suzanne Slade

> Note-
> These books are not required, so we suggest you get these from the library. If they don't have these titles, check out another option your library does have on the topic.

Do: Experimenting with Science

Experiment: When does water change state?

> Caution-
> THIS EXPERIMENT SHOULD NOT BE DONE WITHOUT ADULT SUPERVISION!
> Do not let the students touch the burner! Be sure everyone involved uses protective gear when working near the heat.

Materials

You will need the following items:

- ✓ Cup
- ✓ Ice cubes
- ✓ Pot
- ✓ Thermometer

Introduction

Read the following with your students, while they follow along in their lab journal.

> Addy's father taught the Whitaker girls how to play a game with the states of matter. Remember that there are three states of matter that are easily observed every day on Earth: solid, liquid and gas. Solids have tightly packed molecules with fixed shape and volume, liquids have widely spaced molecules with a fixed volume and gases have independently moving molecules with no fixed shape or volume. There is also a fourth state of matter called plasma, which is rarely seen on Earth. Plasma is created when a gas ionizes at very high temperatures. In this experiment, you are going to look at the points at which water changes state.

Directions

Have the students do the following experiment to observe changes in state.

1. Fill a small pot halfway with ice cubes.
2. Place the pot on a burner and turn the burner on to medium heat.
3. Observe the thermometer as the ice begins to melt and record the temperature once all of the ice melts.
4. Continue to heat the water, observing the temperature on the thermometer as it heats up.
5. Once you begin to see the water boiling and observe the presence of steam, record your last temperature measurement.

6. Turn the burner off and remove the pot from the burner.

Troubleshooting Tips – If the temperature goes way up, but your water is not boiling, make sure that your thermometer is not touching the bottom of the pan. If it is, you will get a false reading. Also, your elevation can affect your results. The melting point and boiling point of water are given at 1 atmosphere of pressure, which is at sea level. If you live above or below sea level, this will affect the temperature at which your ice melts and your water boils. The effect will depend upon how much you are above or below sea level. Generally, the higher your elevation, the higher your temperatures will be.

Results and Explanation

When they are done, read the results and explanation to the students.

> You should have seen the water began to melt at 32° F (0° C) and that it began to boil around 212° F (100° C).

Discuss how the results from their experiment looked like or didn't look like the expected results. Then, read the following explanation to the students.

> The melting point of water is 32° F (0° C), which is the point at which it changes from a solid to a liquid. This is the reason that you began to see the ice melting to liquid water around that temperature. The boiling point of water is 212° F (100° C), which is the point at which it changes from a liquid to a gas. This is the reason that you began to see steam rising from the pot once around that temperature.

(Optional) Hands-on Science Activities

At another point in the week, the students can do one or more of the following optional activities.

- **States of Matter Poster** – Have the students research the four states of matter (gas, solid, liquid, and plasma) and the names of the different phase changes. Next, have them create a poster with the title "States of Matter." Then, have them divide the poster into 4 sections and label each one with a different state of matter. In each box have them share what they have learned from their research in a visually attractive manner (e.g., including pictures). Finally, have the students present a brief report on the states of matter using their poster as a visual aid.
- **Dry Ice Exploration** – Dry ice is the solid form of carbon dioxide. It's a fun chemical to play around with because it readily sublimes at atmospheric pressure. Check out the following article for activity suggestions and a coordinating worksheet.

  https://elementalscience.com/blogs/science-activities/3-science-activities-to-explore-the-chemistry-of-dry-ice

Write: Keeping a Journal

Journaling Sheets

The students should have already recorded what they learned about states of matter on their notes page (LJ p. 21) and completed the experiment sheet (LJ pp. 18-19). On the notes page, the students can also answer the bonus question. Here is an answer for your reference as you discuss this with your students.

- **Bonus Question** – How does the arrangement of particles affect the states of matter? (In solids, particles are packed closely in a fixed, ordered pattern. In liquids, particles are close but can move/slide past each other. In gas, particles are far apart and move in random directions.)

Vocabulary

At some point in the week, have the students look up the following words and add the definitions to the unit vocabulary sheet on LJ pp. 14-16:

- **Boiling Point** – The temperature at which a substance turns from a liquid into a gas.
- **Freezing Point** – The temperature at which a substance turns from a liquid into a solid.
- **Plasma** – The fourth state of matter that exists only at very high temperatures.
- **Sublimation** – Occurs when a substance changes directly from a solid to a gas without changing into a liquid.

Sketch Assignment

Here is the correctly labeled sketch: (LJ p. 20)

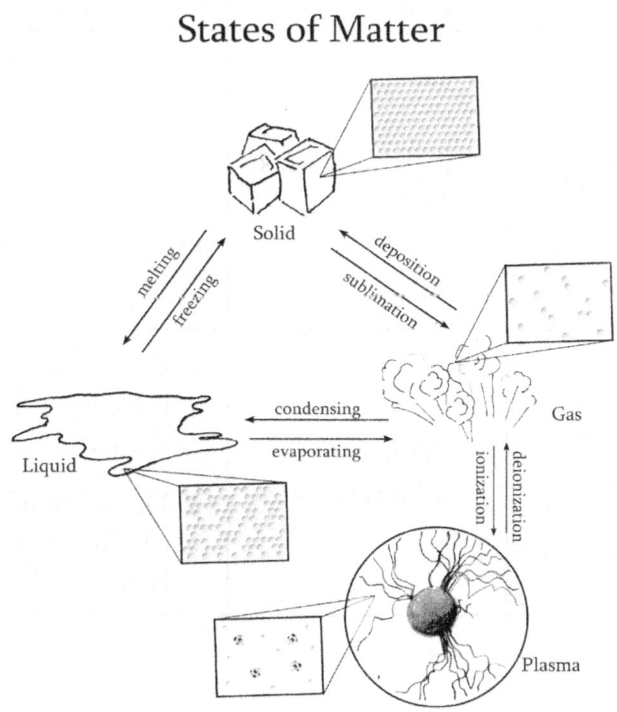

NOTE - The information to label for the sketch assignment can be found on the student's journal sheets.

(Optional) Memory Work

At another point in the week, the students can work on the following memory work - The Gas Laws. For this week, we suggest that you focus on the following:

- **Boyle's Law:** At constant temperature, the volume of a gas is inversely proportional to the pressure.

Lesson 2 Grid Schedule

Supplies Needed	
Experiments	• Wide mouthed jar, String, Pencil, Pipe cleaner, Water, Borax, Scissors
Optional Activities	• Vinegar, Baking soda • Magnifying glass or microscope, Sugar, Salt

Weekly Schedule

	Day 1	Day 2	Day 3	Day 4
Read	❑ Read Chapter 2: "Addy Dug Up A Clue" in *The Route 66 Sticky Note Quest*	❑ Choose one of the encyclopedia readings to assign and discuss what was read using the questions provided	❑ *(Optional)* Choose another one of the encyclopedia readings to assign	❑ *(Optional)* Choose another one of the encyclopedia readings to assign
Do	❑ Do the "Can I make crystals?" experiment	❑ *(Optional)* Make Hot Ice		❑ *(Optional)* Look closer at crystals
Write	❑ Fill out the experiment sheet	❑ Write about solid structures	❑ Color and label the crystal shapes sketch ❑ Define cleavage, conductivity, and lattices	❑ *(Optional)* Work on memorizing the Gas Laws

Lesson 2 List Schedule

Essentials

Read

- O Read Chapter 2: "Addy Dug Up A Clue" in *The Route 66 Sticky Note Quest.*
- O Choose one of the encyclopedia readings to assign and discuss what was read using the questions provided.

Do

- O Do the "Can I make crystals?" experiment. Fill out the experiment sheet as you do the experiment.

Write

- O Write about solid structures.
- O Look up cleavage, conductivity, and lattices. Add the definitions to the unit vocabulary sheet.
- O Label the crystal shapes sketch.

(Optional) Extras

Read

- O Choose another one (or two) of the encyclopedia readings to assign.

Do

- O Make a Hot Ice.
- O Look closer at crystals.

Write

- O Work on memorizing the Gas Laws.

Supplies Needed	
Experiments	• Wide mouthed jar, String, Pencil, Pipe cleaner, Water, Borax, Scissors
Optional Activities	• Vinegar, Baking soda • Magnifying glass or microscope, Sugar, Salt

Lesson 2 Science Focus: Solid Structures

Read: Gathering Information

Key Idea

Crystals have atoms that are arranged in orderly, repeated patterns.

Week's Readings

First, spark an interest by having the students read this week's story opener:

- Read Chapter 2 in *The Route 66 Sticky Note Quest*: "Addy Dug Up A Clue." (Your student will join Addy and the Whitaker family as they go on a hunt at Craters of Diamonds State Park.)

Then follow the story up with some facts. You can choose which resource works best for your students:

- *Usborne Science Encyclopedia*: "Crystals" pp. 90-91
- *Kingfisher Science Encyclopedia:* "Crystals" pp. 168-169
- *Usborne Illustrated Dictionary of Science:* "Crystals" pp. 135-137
- *Everything You Need to Ace Science in One Big Fat Notebook:* "Minerals and Their Uses" pp. 228-230

Discussion Questions

After all that reading, use the following questions to guide your discussion time with the students.

1. What are crystals? (Crystals are solids with regular geometric shapes that can be formed from the regular arrangement of atoms, ions, or molecules.)
2. What are two methods of crystallization? (Two methods of crystallization are allowing a solvent to evaporate and placing a seed crystal in a supersaturated solution.)
3. What causes crystals to form a definite shape? (Crystals form a definite shape because of the arrangement of their atoms or ions.)
4. What are the main crystal shapes? (The five basic shapes of crystals are cubic, tetragonal, monoclinic, triclinic [or rhombohedral], and hexagonal.)
5. What is polymorphism in crystals? (Polymorphism in crystals is when a substance can have two or more different crystal shapes.)
6. What are the four main types of lattice structures? Describe each one. (The four main types of lattice structures are giant atomic, giant ionic, giant metallic, and molecular lattices. Giant atomic lattices are crystal lattices held together with covalent bonding. Giant ionic lattices are crystal lattices held together with ionic bonding. Giant metallic lattices are crystal lattices held together by metallic bonding with delocalized electrons. Molecular lattices are crystal lattices held together by weak intermolecular forces.)

After the discussion time, have the students write a list of facts or a paragraph about what they read. Their writings should include at least a few facts from above.

(Optional) Reading Selections

At another point in the week, the students can check out these optional books:

📖 *Gems for Kids: A Junior Scientist's Guide to Mineral Crystals and Other Natural Treasures* by Ashley Hall and Lee Hall

📖 *The Rock and Gem Book: And Other Treasures of the Natural World* by DK and Smithsonian Institution

Do: Experimenting with Science

Experiment: Can I make crystals?

Materials

You will need the following items:

- ✓ Wide mouthed jar
- ✓ String
- ✓ Pencil
- ✓ Pipe cleaner
- ✓ Water
- ✓ Borax
- ✓ Scissors

> Note–
>
> You can find Borax in the laundry aisle of the local grocery store. Be sure to buy the one labeled laundry booster, not the soap that includes Borax.

Introduction

Read the following with your students, while they follow along in their lab journal.

Addy and her family learn about and dig up different minerals during their visit to a state park. Minerals have crystal structures in all shapes and sizes, but each is based on the regularly repeating pattern of the atoms, molecules, or ions that make it up. They are very hard, solid structures, and we use them in a variety of ways as we go about our daily lives. For example, the salt you use on your food and the lead in your pencil are both made up of crystals. In this experiment, you are going to make your own crystals.

Directions

Have the students do the following experiment to use a leaf to learn about a local tree under your supervision.

1. Fill your jar with water, pour that water into a pot, and set it on a burner to boil. Make sure to record how many cups (or mL) of water it takes to fill your jar.
2. Form a shape with the pipe cleaner—this can be as simple or as complex as you wish, but make sure it will fit through the opening of your jar.
3. Next, tie a string to the shape and then tie the other end of the string to a pencil.
4. Set your jar in the sink, put on safety glasses, and use a hot mitt to slowly add boiling water until it almost fills the jar.
5. Then, add the Borax one TBSP (25 g) at a time, stirring each time until the Borax is

> Troubleshooting Tips–
>
> You want the pencil to be able to rest on the edge of your jar without having your shape touch the sides or bottom of the jar.

dissolved. (**Note**—You want to add about 3 TBSP (75 g) of Borax for every cup (240 mL) of water you've added. Be sure to record how many tablespoons you added on your experiment sheet.)

6. Hang the shape in the jar so that it is completely covered by the liquid.
7. Allow the jar to sit undisturbed overnight.
8. In the morning, take out your shape and observe what has happened.

Results and Explanation

When they are done, read the results and explanation to the students.

> You should have seen the pipe cleaner creation is completely coated with crystals. The crystals should be white, making their shape look like it is covered in ice or snow.

Discuss how the results from their experiment looked like or didn't look like the expected results. Then, read the following explanation to the students.

> You super-saturated the water with Borax crystals because hot water can hold more crystals than cold water. This is due to the fact that the hot water molecules are spread farther apart, which creates more room into which the Borax molecules can dissolve. As the water cooled, the molecules came closer together, causing the Borax molecules to be forced out of the solution and form solid crystals once again. The pipe cleaner gave the molecules an easy place to attach, which is why it was coated with crystals.

(Optional) Hands-on Science Activities

At another point in the week, the students can do one or more of the following optional activities.

- **Hot Ice** – This experiment is very cool when it works, but it's a bit temperamental. You will need vinegar and baking soda for the experiment. Directions and troubleshooting tips can be found at:

  https://frugalfun4boys.com/hot-ice-science-experiment/

- **Looking at Crystals** – Have the students use a magnifying glass or microscope to examine sugar and salt crystals. You can also have them examine one of their Borax crystals from the experiment. Talk about the different crystal systems (or shapes) that the students see. (Borax and sugar are both monoclinic, while salt is cubic.)

Write: Keeping a Journal

Journaling Sheets

The students should have already recorded what they learned about solid structures on their notes page (LJ p. 27) and completed the experiment sheet (LJ pp. 24-25). On the notes page, the students can also answer the bonus question. Here is an answer for your reference as you discuss this with your students.

- **Bonus Question** – Would the way a crystal is formed affect its shape? (Yes, the way a crystal forms does affect its shape. For example, if a crystal forms by a liquid cooling, it will typically yield smaller, more numerous crystals due to rapid saturation and decreased movement of the atoms. But if a crystal forms by a liquid evaporating, it can allow for larger, more orderly crystals because it increases concentration gradually without significantly lowering the temperature.)

Vocabulary

At some point in the week, have the students look up the following words and add the definitions to the unit vocabulary sheet on LJ pp. 14-16:

- **Cleavage** – The splitting of a crystal along a certain plane.
- **Conductivity** – The measure of a substance's ability to conduct heat or electricity.
- **Lattices** – The regular arrangement of repeating patterns of atoms, molecules, or ions in a solid structure.

Sketch Assignment

Here is the correctly labeled sketch: (LJ p. 26)

Crystal Shapes

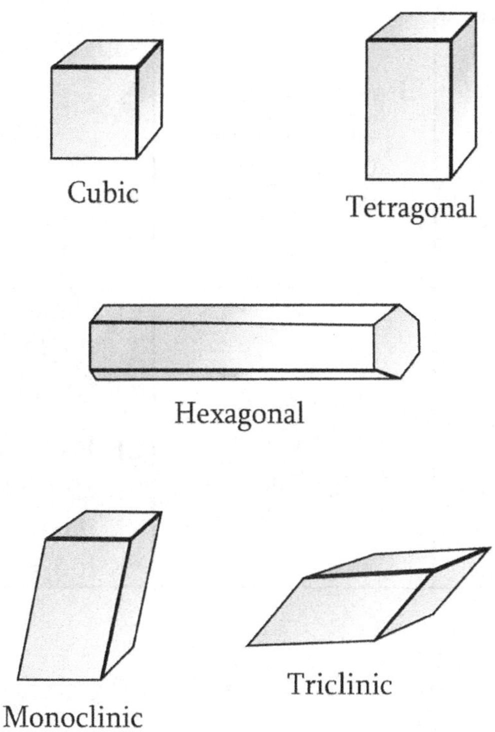

(Optional) Memory Work

At another point in the week, the students can work on the following memory work - the Gas Laws. For this week, we suggest that you focus on the following:

- **Charles' Law:** At constant pressure, the volume of a gas is directly proportional to the temperature. (This law is also known as the Law of Volumes.)

Lesson 3 Grid Schedule

Supplies Needed	
Experiments	• 4 Clear cups, Eye dropper, Table salt, Food coloring, Water
Optional Activities	• Honey, Corn syrup, Dish soap, Water, Vegetable oil, Rubbing alcohol, Lamp oil • Shallow bowl, Needle, Dish soap

Weekly Schedule

	Day 1	Day 2	Day 3	Day 4
Read	❑ Read Chapter 3: "Addy Shared the Sticky Note" in *The Route 66 Sticky Note Quest*	❑ Choose one of the encyclopedia readings to assign and discuss what was read using the questions provided	❑ *(Optional)* Choose another one of the encyclopedia readings to assign	❑ *(Optional)* Choose another one of the encyclopedia readings to assign
Do	❑ Do the "How does the addition of salt affect the density of water?" experiment	❑ *(Optional)* Make a Density Column		❑ *(Optional)* Explore surface tension
Write	❑ Fill out the experiment sheet	❑ Write about molecular properties	❑ Color and label the Density Column ❑ Define density, elasticity, and intermolecular forces	❑ *(Optional)* Work on memorizing the Gas Laws

Lesson 3 List Schedule

Essentials

Read

- O Read Chapter 3: "Addy Shared the Sticky Note" in *The Route 66 Sticky Note Quest.*
- O Choose one of the encyclopedia readings to assign and discuss what was read using the questions provided.

Do

- O Do the "How does the addition of salt affect the density of water?" experiment. Fill out the experiment sheet as you do the experiment.

Write

- O Write about molecular properties.
- O Look up density, elasticity, and intermolecular forces. Add the definitions to the unit vocabulary sheet.
- O Label the Density Column sketch.

(Optional) Extras

Read

- O Choose another one (or two) of the encyclopedia readings to assign.

Do

- O Make a Density Column.
- O Explore surface tension.

Write

- O Work on memorizing the Gas Laws.

Supplies Needed	
Experiments	• 4 Clear cups, Eye dropper, Table salt, Food coloring, Water
Optional Activities	• Honey, Corn syrup, Dish soap, Water, Vegetable oil, Rubbing alcohol, Lamp oil • Shallow bowl, Needle, Dish soap

Lesson 3 Science Focus: Molecular Properties

Read: Gathering Information

> **Key Idea**
>
> **Density is a property of matter that relates mass and volume.**

Week's Readings

First, spark an interest by having the students read this week's story opener:

- Read Chapter 3 in *The Route 66 Sticky Note Quest*: "Addy Shared the Sticky Note." (Your student will follow Addy as the Whitaker family travels toward the start of Route 66.)

Then follow the story up with some facts. You can choose which resource works best for your students:

- *Usborne Science Encyclopedia*: "How Liquids Behave" pp. 20-21
- *Kingfisher Science Encyclopedia*: (no pages scheduled)
- *Usborne Illustrated Dictionary of Science:* "Molecular Properties" pp. 22-23, "Density" p. 24
- *Everything You Need to Ace Science in One Big Fat Notebook:* "Buoyancy and Density" p. 116

Discussion Questions

After all that reading, use the following questions to guide your discussion time with the students.

1. How is elasticity related to intermolecular forces? (When a material is stretched or compressed, the molecules in it are attracted or repelled to each other. So when the stretch or compression is removed from the material, the intermolecular forces cause it to return to the way it was.)
2. What does Hooke's Law say? (Hooke's Law says that when a distorting force is applied to an object, the strain is proportional to the stress.)
3. What is surface tension? (Surface tension is the skin-like properties of a liquid's surface due to intermolecular forces.) Adhesion? (Adhesion is the intermolecular force of attraction between molecules within different substances.) Cohesion? (Cohesion is the intermolecular force of attraction between molecules within the same substance.)
4. What is specific gravity? (Specific gravity, also known as relative density, is the density of a substance relative to the density of water.)
5. What does a hydrometer measure? (A hydrometer measures the density of a liquid.)

After the discussion time, have the students write a list of facts or a paragraph about what they read. Their writings should include at least a few facts from above.

(Optional) Reading Selections

At another point in the week, the students can check out these optional books:

- *Will I Sink or Float? Explaining Mass, Volume, Density and the Importance of SI Units* by Baby Professor
- *Density: The Hatgrams and McCubes: Another Hare-Brain Science Tale* by John A. Honeycutt

Do: Experimenting with Science

Experiment: How does the addition of salt affect the density of water?

Materials

You will need the following items:

- ✓ 4 Clear cups
- ✓ Eye dropper
- ✓ Table salt
- ✓ Food coloring
- ✓ Water

Introduction

Read the following with your students, while they follow along in their lab journal.

> Addy finds a note with a riddle and the answer to it was density. Density is a property of matter that relates mass and volume. Mass tells you how much matter is in something, while volume tells you how much space something occupies. So in other words, density measures how much matter of a particular substance will occupy a given space. When comparing two different solutions, a more dense solution will sink to the bottom, while a less dense solution will float to the top. In this week's experiment, you are going to use the principle of density to see how salt affects the density of water.

Directions

Have the students do the following experiment to observe density under your supervision.

Prepare the Solutions

1. Begin by putting 1 cup (240 mL) of room temperature water into each of the 4 bowls.
    - For the yellow solution, add 10 drops of yellow food coloring to one of the bowls.
    - For the red solution, add 10 drops of red food coloring and 2 TBSP (34.2 g) of salt to another bowl.
    - For the blue solution, add 10 drops of blue food coloring and ¼ cup (68.4 g) of salt to another bowl.
    - For the green solution, add 10 drops of green food coloring and 6 TBSP (102.6 g) of salt to last bowl.
2. Stir each one to make sure that the salt is thoroughly dissolved.

Use the Solutions

1. Label each of your cups #1-4. Fill cup #1 with 1 cup (240 mL) of the yellow solution, cup #2 with 1 cup (240 mL) of the red solution, cup #3 with 1 cup (240 mL) of the blue solution, and cup #4 with 1 cup (240 mL) of the green solution.
2. Using an eye dropper, draw up about 1-2 tsp (5-10 mL) of the yellow solution and very slowly add it to cup #2. (**Note**—When you add the yellow solution, you need to make sure the tip of your eyedropper is just below the surface of the solution in the cup.) Wait 1 minute and record your observations.

3. Next, using an eye dropper, draw up about 1-2 tsp (5-10 mL) of the yellow solution and very slowly add it to cup #3. Wait 1 minute and record your observations.
4. Then, using an eye dropper, draw up about 1-2 tsp (5-10 mL) of the yellow solution and very slowly add it to cup #4. Wait 1 minute and record your observations.
5. Pour out the solutions from each of the used cups and refill them with 1 cup of each of the colors as before.
6. Repeat steps 3 through 6 with the red, blue, and green solutions. (**Note**—Each time you will skip the cup with the same color solution because we already know that they have the same concentration of salt.)
7. Record your observations and results.

Results and Explanation

When they are done, check the results using the chart below and read the explanation to the students.

	Added to the Yellow Solution	Added to the Red Solution	Added to the Blue Solution	Added to the Green Solution
Yellow Solution		Floated	Floated	Floated
Red Solution	Sank		Floated	Floated
Blue Solution	Sank	Sank		Floated
Green Solution	Sank	Sank	Sank	

Discuss how the results from their experiment looked like or didn't look like the expected results. Then, read the following explanation to the students.

> The addition of salt to the water increases the overall mass of the solution, but minimally affects the volume. This means that the density will increase. The more salt you add to the water, the more dense the solution will become. This is the reason why you saw that the yellow liquid always floated to the top of each cup, while the green liquid always sank to the bottom.

(Optional) Hands-on Science Activities

At another point in the week, the students can do one or more of the following optional activities.

- **Density Column** – Have the students make a density column. You will need ½ cup (120 mL) of honey, corn syrup, dish soap, water, vegetable oil, rubbing alcohol, and lamp oil each in separate cups. Add a different shade of food coloring to each of the liquids, except the honey and vegetable oil. (**Note**—If your dish soap is already colored, you do not need to add a color to that.) Next, pour the honey into a cylindrical container. Then, slowly add the corn syrup on top. You want to pour the liquid slowly down the side of the container to minimize the bubbles and give you the best results. Follow with the dish soap, water, vegetable oil, alcohol, and finally the lamp oil. You should have seven separate liquid layers on top of each other. Each layer has the same volume, but different densities.
- **Surface Tension** – Have the students play with surface tension. You will need a shallow bowl, a needle, and dish soap. Fill the bowl about halfway with water. Gently place the needle on the surface without touching the water. Observe what happens to the surrounding water.

(If you look closely, the student should see that the needle looks like it is resting on the "skin" of the water formed by surface tension.) Add a drop of dish soap near the needle and observe what happens to the needle. (The needle should sink to the bottom because the soap molecules break the weak intermolecular forces that form the surface tension "skin" of the water.)

Write: Keeping a Journal

Journaling Sheets

The students should have already recorded what they learned about molecular properties and density on their notes page (LJ p. 33) and completed the experiment sheet (LJ pp. 30-31). There is no bonus question this week.

Vocabulary

At some point in the week, have the students look up the following words and add the definitions to the unit vocabulary sheet on LJ pp. 14-16:

- **Density** – A measure of the amount of matter in a substance compared to its volume.
- **Elasticity** – The ability of a substance to stretch and then return to its original shape.
- **Intermolecular forces** – The forces that exist between molecules, which explain properties like elasticity, surface tension, and viscosity.

Sketch

Here is the correctly labeled sketch. (LJ p. 32)

Density Column

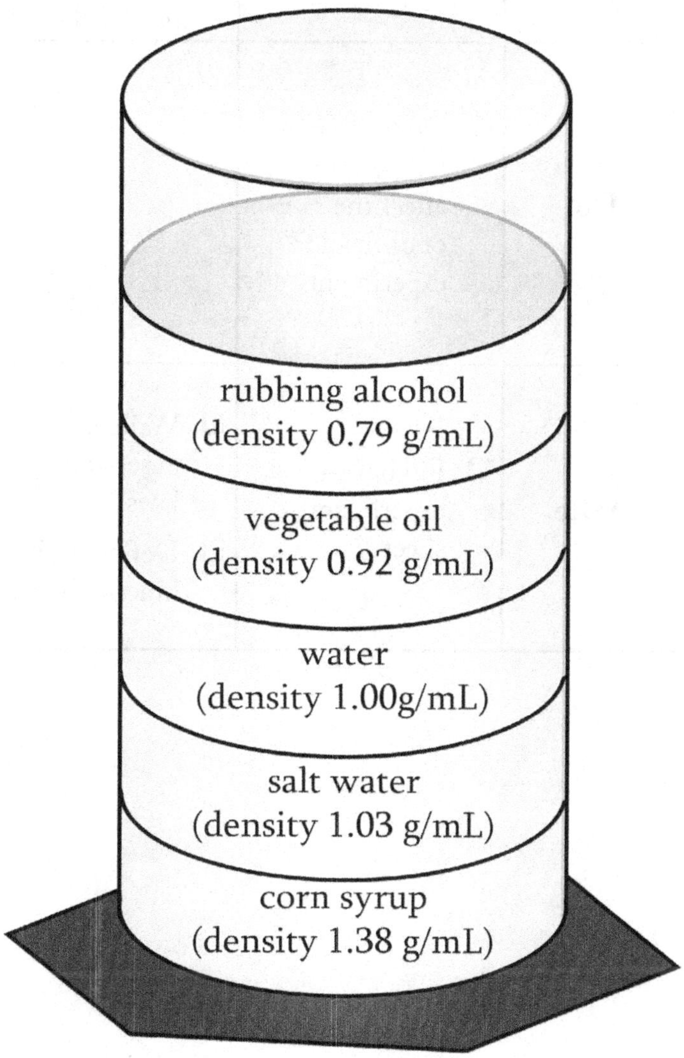

(Optional) Memory Work

At another point in the week, the students can work on the following memory work - the Gas Laws. For this week, we suggest that you focus on the following:

- **Gay-Lussac's Law:** At constant volume and mass, the pressure of a gas is directly proportional to its absolute temperature.

Lesson 4 Grid Schedule

Supplies Needed	
Experiments	• 2 Cups, Apple juice, Timer, Partner
Optional Activities	• Marshmallow, Large needleless syringe • 2 Balloons, Water, Ice

Weekly Schedule				
	Day 1	**Day 2**	**Day 3**	**Day 4**
Read	❑ Read Chapter 4: "Addy Found a Box" in *The Route 66 Sticky Note Quest*	❑ Choose one of the encyclopedia readings to assign and discuss what was read using the questions provided	❑ *(Optional)* Choose another one of the encyclopedia readings to assign	❑ *(Optional)* Choose another one of the encyclopedia readings to assign
Do	❑ Do the "Does temperature affect the rate of diffusion?" experiment		❑ *(Optional)* Test Boyle's Law	❑ *(Optional)* Test Charles's Law
Write	❑ Fill out the experiment sheet	❑ Write about gases and the gas laws ❑ Define diffusion, kinetic theory	❑ Color and label the Gas Laws sketch ❑ Define pressure, temperature, and volume	❑ *(Optional)* Take the Unit 1 Test

Lesson 4 List Schedule

Essentials

Read

- O Read Chapter 4: "Addy Found a Box" in *The Route 66 Sticky Note Quest.*
- O Choose one of the encyclopedia readings to assign and discuss what was read using the questions provided.

Do

- O Do the "Does temperature affect the rate of diffusion?" experiment. Fill out the experiment sheet as you do the experiment.

Write

- O Write about gases and the gas laws.
- O Look up diffusion, kinetic theory, pressure, temperature, and volume. Add the definitions to the unit vocabulary sheet.
- O Label the Gas Laws sketch.

(Optional) Extras

Read

- O Choose another one (or two) of the encyclopedia readings to assign.

Do

- O Test Boyle's Law.
- O Test Charles's Law.

Write

- O Take the Unit 1 Test.

Supplies Needed	
Experiments	• 2 Cups, Apple juice, Timer, Partner
Optional Activities	• Marshmallow, Large needleless syringe • 2 Balloons, Water, Ice

Lesson 4 Science Focus: Gas Laws

Read: Gathering Information

Key Idea

An ideal gas behaves in an ideal way.

Week's Readings

First, spark an interest by having the students read this week's story opener:

- Read Chapter 4 in *The Route 66 Sticky Note Quest*: "Addy Found a Box." (Your student will follow the Whitaker and Bell families as they tour Chicago. On the way back to the RV, Addy finds a box with another sticky note.)

Then follow the story up with some facts. You can choose which resource works best for your students:

- *Usborne Science Encyclopedia*: "How Gases Behave" pp. 22-23
- *Kingfisher Science Encyclopedia*: "Pressure" p. 311 (You will be learning more about pressure in lesson 15, so you might want to choose to save these pages.)
- *Usborne Illustrated Dictionary of Science:* "Kinetic Theory" p. 123, "Gas Laws" pp. 142-143
- *Everything You Need to Ace Science in One Big Fat Notebook:* "Pressure" pp. 87-88 (You will be learning more about pressure in lesson 15, so you might want to choose to save these pages.)

Discussion Questions

After all that reading, use the following questions to guide your discussion time with the students.

1. What are the three gas laws? (Boyle's Law says that at constant temperature, the volume of a gas is inversely proportional to the pressure. The Law of Volumes says that at constant pressure, the volume of a gas is directly proportional to the temperature. The Third Gas Law says that at constant volume, the pressure is directly proportional to the temperature.)
2. What is an ideal gas? (An ideal gas is one that behaves in an ideal way, its molecules have no volume or intermolecular forces and move in straight lines, losing no energy when they collide.)
3. What is Dalton's law of partial pressure? (Dalton's law of partial pressure says that the sum of the partial pressure of each gas in a mixture is equal to the total pressure.)
4. What does Graham's law of diffusion tell us? (Graham's law of diffusion tells us that the faster a gas diffuses, the lower its density, and vice versa.)
5. What is stp? (Stp is an abbreviation for standard temperature and pressure.)
6. What is Avogadro's Law? (Avogadro's law states that equal volumes of gases at the same temperature and pressure contain the same number of molecules.)

After the discussion time, have the students write a list of facts or a paragraph about what they read. Their writings should include at least a few facts from above.

(Optional) Reading Selections

At another point in the week, the students can check out these optional books:

- *Gases* by William Rice

📖 *Farts, Fumes, and Other Gases (Where Does It Go?)* by Helen Greathead

Do: Experimenting with Science

Experiment: Does temperature affect the rate of diffusion?

Materials

You will need the following items:

- ✓ 2 Cups
- ✓ Apple juice
- ✓ Timer
- ✓ Partner

CAUTION-
THIS EXPERIMENT SHOULD NOT BE DONE WITHOUT ADULT SUPERVISION!
The apple juice could be very hot when removing it from the microwave. Make sure to use protective gear.

Introduction

Read the following with your students, while they follow along in their lab journal.

> Peter explains the gas laws to Addy and her sister. The particles in a substance are constantly in motion. In a solid, these particles are tightly packed, so they can only vibrate or shake. In a liquid, these particles can move more freely, but they still remain relatively close. Particles in a gas are widely spaced and can move very quickly. When the particles in a substance move from one corner of a room to the other, they diffuse, or spread out. In this week's experiment, you are going to examine how temperature affects the rate of diffusion.

Directions

Have the students test how temperature affects diffusion under your supervision.

1. Fill 2 cups with apple juice, cover each of them with plastic wrap, and leave them on the counter.
2. After one hour, take one of the cups from the counter and have your partner stand in the opposite corner of the room.
3. Remove the plastic wrap from the cup and time how long it takes for them to smell the apple juice.
4. Record that time on your experiment sheet. (**Note**—If your partner doesn't smell it after 10 minutes, end the test and note the results on the experiment sheet.)
5. Next, take the other cup from the counter and microwave it on high for 1-2 minutes, until the juice is warm, but not too hot to touch.
6. Have your partner stand in the opposite corner of the room and remove the plastic wrap from the cup.
7. Time how long it takes for them to smell the apple juice and record that time on your experiment sheet. (**Note**—If your partner doesn't smell it after 10 minutes, end the test and note the results on the experiment sheet.)

8. Record your observations and results before drawing conclusions.

Results and Explanation

When they are done, read the results and explanation to the students.

> You should have seen the hot apple juice diffused through the room much quicker than the room temperature apple juice. In fact, you may have seen that your partner never smelled the room temperature apple juice.

Discuss how the results from their experiment looked like or didn't look like the expected results. Then, read the following explanation to the students.

> As the particles in the juice heat up, their energy of motion increases. This means that they are able to escape from the glass and travel around the room quicker than the room temperature molecules. This is known as the kinetic theory, which states that as the temperature rises, the particles move faster, and therefore they take up more space.

(Optional) Hands-on Science Activities

At another point in the week, the students can do one or more of the following optional activities.

- **Testing Boyle's Law** – You will need a marshmallow and a large needleless syringe. Begin by removing the plunger and putting the marshmallow into the syringe. Replace the plunger and push it down until it is almost touching the marshmallow. Then, cover the tip of the syringe with your finger to create a vacuum and pull the plunger up. (You should see that the marshmallow expands because as the volume increases, the pressure on the marshmallow decreases.)
- **Testing Charles's Law** – You will need 2 balloons, water, and ice. Blow each of the balloons up to the same size. Place one on the counter and one in a bucket of ice water. Wait for 10 minutes and take the balloon out of the ice water. Compare the two balloons side by side. (You should see that the one that was in the ice water has decreased in size because as the temperature decreases, the volume of a gas will also decrease.)

Write: Keeping a Journal

Journaling Sheets

The students should have already recorded what they learned about gases and the gas laws on their notes page (LJ p. 39) and completed the experiment sheet (LJ pp. 36-37). On the notes page, the students can also answer the bonus question. Here is an answer for your reference as you discuss this with your students.

- **Bonus Question** – How do scuba divers take advantage of Boyle's Law? (Scuba divers are taught to breathe continuously and exhale during ascent to allow expanding air to escape their lungs because as they rise, the pressure decreases and the air expands. Divers use air from their tanks to add volume to masks, BCDs (Buoyancy Compensators), and ears during descent (as air compresses) and allow it to escape during ascent (as air expands).)

Vocabulary

At some point in the week, have the students look up the following words and add the definitions to the unit vocabulary sheet on LJ pp. 14-16:

- **Diffusion** – The spreading out of a gas to fill the available space of the container it is in.
- **Kinetic Theory** – The theory that states that as the temperature rises, particles move faster and therefore take up more space.
- **Pressure** – The amount of force that pushes on a given area.
- **Temperature** – A measure of the heat energy that a substance contains.
- **Volume** – A measurement of the space occupied by a substance.

Sketch

Here is the correctly labeled sketch: (LJ p. 38)

Gas Laws

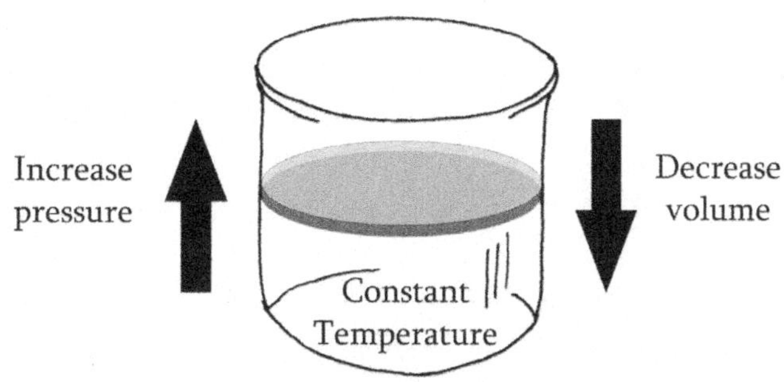

Boyle's Law

Charles's Law
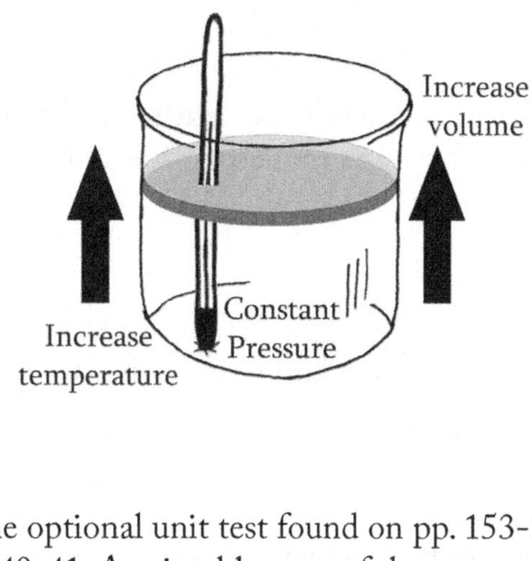

(Optional) Matter Test

The students can complete the optional unit test found on pp. 153-155 in the Appendix. The answers for this test are found on pp. 40-41. A printable copy of the test can be downloaded for free from:

💻 https://elementalscience.com/blogs/resources/sq200

Matter Test Answers

Vocabulary Matching (2 points each)

> Note -
> If you need a digital copy of the unit test, you can download one for free from this program's resource page. Just head to:
> https://elementalscience.com/blogs/resources/sq200

1. E
2. I
3. B
4. D
5. N
6. F
7. K
8. A
9. C
10. O
11. M
12. H
13. G
14. J
15. L

True and False (2 points each)

1. True
2. False (The boiling point is the point at which a liquid changes into a gas and the melting point is the point at which a solid changes into a liquid.)
3. True
4. True
5. False (Adhesion is the intermolecular force of attraction between molecules within different substances, while cohesion is the intermolecular force of attraction between molecules within the same substance.)
6. True
7. True
8. False (Stp is an abbreviation for standard temperature and pressure.)

Short Answer

1. (7 points) The three states of matter that are typically found on Earth are solid, liquid, and gas. A solid is a state of matter that has definite shape and volume. A liquid is a state of matter that has fixed volume, but no definite shape. A gas is a state of matter that has no fixed

volume or shape.

2. (2 points) Crystals form a definite shape because of the arrangement of their atoms or ions.
3. (2 points) Surface tension is the skin-like properties of a liquid's surface due to intermolecular forces.
4. (3 points) At constant temperature, the volume of a gas is inversely proportional to the pressure.
5. (3 points) At constant pressure, the volume of a gas is proportional to the temperature.

Total points: 75

SciQuest

Addy's Adventure Teacher Notes

Solutions and Reactions Unit

What you'll study in this unit-

Memory Work-

Law of Constant Composition and Conservation of Mass (LJ p. 43)

Lesson 5 Grid Schedule

Supplies Needed	
Experiments	• Bag of multi-colored marshmallows, Toothpicks
Optional Activities	• More toothpicks and marshmallows • Water, Oil, Juice, Alcohol, Food coloring

Weekly Schedule				
	Day 1	Day 2	Day 3	Day 4
Read	❑ Read Chapter 5: "Addy Rode on Route 66" in *The Route 66 Sticky Note Quest*	❑ Choose one of the encyclopedia readings to assign and discuss what was read using the questions provided	❑ *(Optional)* Choose another one of the encyclopedia readings to assign	❑ *(Optional)* Read one or both of the optional reading selections
Do	❑ Do the "Marshmallow Molecules" activity	❑ *(Optional)* Make a polyethylene chain with marshmallows.	❑ *(Optional)* Test for miscible or immiscible liquids	
Write	❑ Fill out the experiment sheet	❑ Write about compounds and mixtures	❑ Define immiscible, miscible, mixture, and molecule	❑ *(Optional)* Work on memorizing the Law of Constant Composition

Lesson 5 List Schedule

Essentials

Read

- O Read Chapter 5: "Addy Rode on Route 66" in *The Route 66 Sticky Note Quest*.
- O Choose one (or more) of the encyclopedia readings to assign and discuss what was read using the questions provided.

Do

- O Do the "Marshmallow Molecules" activity. Fill out the experiment sheet as you do the experiment.

Write

- O Write about compounds and mixtures.
- O Look up immiscible, miscible, mixture, and molecule. Add the definitions to the unit vocabulary sheet.

(Optional) Extras

Read

- O Read one or both of the optional reading selections.

Do

- O Make a polyethylene chain with marshmallows.
- O Test for miscible or immiscible liquids.

Write

- O Work on memorizing the Law of Constant Composition.

Supplies Needed	
Experiments	• Bag of multi-colored marshmallows, Toothpicks
Optional Activities	• More toothpicks and marshmallows • Water, Oil, Juice, Alcohol, Food coloring

Lesson 5 Science Focus: Molecules

Read: Gathering Information

> **Key Idea**
> **Molecules are formed when two or more atoms join together.**

Week's Readings

First, spark an interest by having the students read this week's story opener:

- Read Chapter 5 in *The Route 66 Sticky Note Quest*: "Addy Participated in the Whitaker Pilgrimage." (Your student will watch Addy and her sisters try to figure out the clue that the box revealed.)

Then follow the story up with some facts. You can choose which resource works best for your students:

- *Usborne Science Encyclopedia*: "Molecules" pp. 14-15
- *Kingfisher Science Encyclopedia:* "Chemical Compounds" pp. 164-165
- *Usborne Illustrated Dictionary of Science:* "Elements, Compounds, and Mixtures (except the section on kinetic theory) " pp. 122-123
- *Everything You Need to Ace Science in One Big Fat Notebook:* "Molecules and Compounds" pp. 78-80, "Mixture" p. 84

Discussion Questions

After the reading, use the following questions to guide your discussion time with the students.

1. What is synthesis? (Synthesis is when chemists build larger and more useful compounds from elements or smaller compounds.)
2. How are compounds and mixtures alike? (Compounds and mixtures are both a combination of different elements.) How are they different? (Compounds have their own unique chemical properties different from the elements that make them up, while compounds or elements in a mixture still retain their original chemical properties. A mixture can be easily separated, while compounds must be chemically separated.)
3. What does homogenous mean? Give an example of a homogenous substance. (Homogenous means that a substance has all the particles in the same phase. In other words, the chemical and physical properties are the same throughout the substance. A solution is an example of a homogenous substance.)
4. What does heterogenous mean? Give an example of a heterogenous substance. (Heterogenous means that a substance has particles that are in different phases. In other words, the chemical and physical properties can be different throughout the substance. A suspension is an example of a heterogenous substance.)
5. How do chemists define the purity of a substance? (Chemists say that a substance is pure when it contains only one type of element or compound.)

After the discussion time, have the students write a list of facts or a paragraph about what they read. Their writings should include at least a few facts from above.

(Optional) Reading Selections

At another point in the week, the students can check out these optional books:

- *Reactions: An Illustrated Exploration of Elements, Molecules, and Change in the Universe* by Theodore Gray (This book is scheduled as an optional reading selection throughout this unit.)
- *Atoms, Molecules, And Compounds* by Jenny Karpelenia

Do: Experimenting with Science

Activity: Marshmallow Molecules

Materials

You will need the following items:

- ✓ Bag of multi-colored marshmallows
- ✓ Toothpicks

Introduction

Read the following with your students, while they follow along in their lab journal.

Addy and her sisters hypothesized that the letters on the clues formed molecules. Molecules are formed when two or more atoms are held together by a chemical bond. These molecules form the millions of substances that surround us in daily life, such as water and table salt. The atoms tend to follow a set of rules when they form molecules. In this experiment, you are going to make marshmallow representations of several common molecules.

Directions

Have the students do the following experiment to make models of molecules under your supervision.

1. Choose a color of marshmallow to represent each of the following atoms: oxygen, nitrogen, hydrogen, and carbon.
2. When you make your molecules you must follow these rules:
   - Oxygen prefers to bond twice, nitrogen prefers to bond three times, hydrogen prefers to only bond once, and carbon prefers to bond four times.
   - Each atom must have its preferred number of bonds to form a stable molecule and you must only create stable molecules.
   - All of nitrogen's bonds should point down and all of carbon's bonds need to be opposite from each other (except in the case of a multiple bond).
3. Make the following molecules from your marshmallows:
   - NH_3 (ammonia)
   - H_2O (water)
   - CH4 (methane)
   - CO_2 (carbon dioxide)
   - C_2H_5OH(ethanol)

   (**Hint**—You can make multiple bonds between the atoms.)
4. Draw a quick sketch of each of your molecules after you finish assembling them.

Results and Explanation

This experiment is meant to be a demonstration of how molecules are formed so that the concepts will be solidified in your students' mind. The pictures below will give you an idea of what their creations should look like:

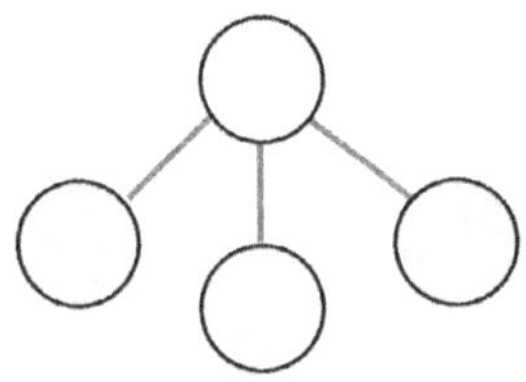

NH_3 (ammonia)

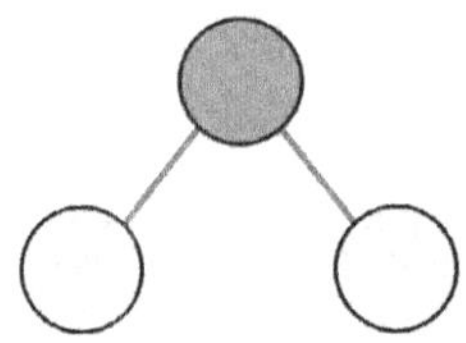

H_20 (water)

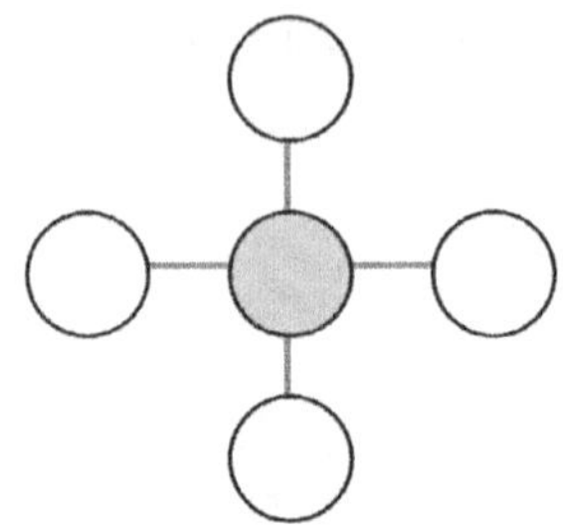

CH_4 (methane)

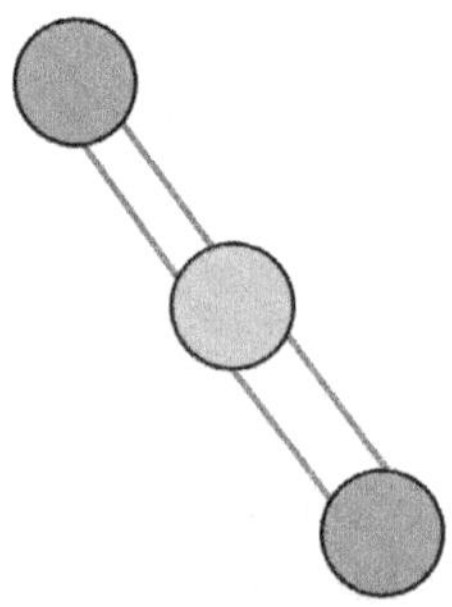

CO_2 (carbon dioxide)

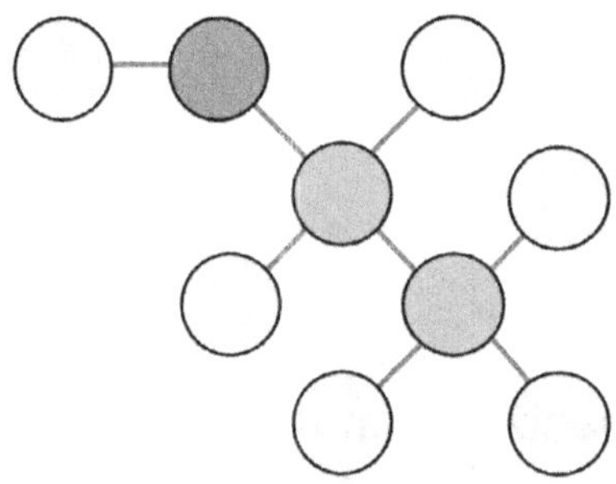

C_2H_5OH(ethanol)

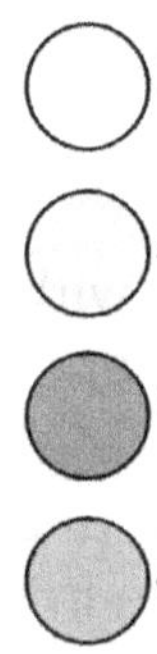

(Optional) Hands-on Science Activities

At another point in the week, the students can do one or more of the following optional activities.

- **Polyethylene Molecule** – Have the students see how long of a polyethylene chain they can make, using the same marshmallows, toothpicks, and rules set forth in the experiment. (Hint—Polyethylene is composed of only carbon and hydrogen.)
- **Miscible or Immiscible** – Have the students gather water, oil, juice, alcohol and food coloring from your kitchen pantry or fridge. Then, have them mix each of the liquids with one of the other liquids to determine if they are miscible or immiscible. (This concept will be touched on again next week.)

Write: Keeping a Journal

Journaling Sheets

The students should have already recorded what they learned about compounds and mixtures on their notes page (LJ p. 51) and completed the experiment sheet (LJ pp. 48-49). On the notes page, the students can also answer the bonus question. Here is an answer for your reference as you discuss this with your students.

- **Bonus Question** – What is the difference between compounds and molecules? (Molecules are formed when two or more atoms join together. Compounds are formed when two or more elements join together. For example H_2 (hydrogen gas) is a molecule because two atoms of hydrogen are joined together. However, since there is only one type of element present, H_2 is not a compound. On the other hand, H_2O (water) is a molecule because the three atoms, one oxygen atom and two hydrogen atoms, have been joined together to form it. It is also a compound because it contains two different elements, hydrogen and oxygen. So, all compounds are molecules, but not all molecules are compounds.)

Vocabulary

At some point in the week, have the students look up the following words and add the definitions to the unit vocabulary sheet on LJ pp. 44-46:

- **Immiscible** – Liquids that cannot be blended together.
- **Miscible** – Liquids that can be blended together.
- **Mixture** – A combination of two or more elements or compounds that are not chemically combined.
- **Molecule** – A substance that is formed when two or more atoms chemically join together.

Sketch Assignment

There is no sketch for this week.

(Optional) Memory Work

At another point in the week, the students can work on the following memory work - The Law of Constant Composition.

- **Law of Constant Composition** – A pure compound always contains the same elements in the same proportions.

Lesson 6 Grid Schedule

Supplies Needed	
Experiments	• 5 Clear cups (or beakers), 5 Plastic spoons, Sugar, Salt, Baking soda, Flour, Petroleum jelly, Water, Vegetable oil, Tablespoon
Optional Activities	• Glass, Tap water, Dish soap

Weekly Schedule

	Day 1	Day 2	Day 3	Day 4
Read	❑ Read Chapter 6: "Addy Walked St. Louis" in *The Route 66 Sticky Note Quest*	❑ Choose one of the encyclopedia readings to assign and discuss what was read using the questions provided	❑ *(Optional)* Choose another one of the encyclopedia readings to assign	❑ *(Optional)* Choose another one of the encyclopedia readings to assign
Do	❑ Do the "Is it polar or nonpolar?" experiment	❑ *(Optional)* Work on a research report about polar and nonpolar molecules	❑ *(Optional)* Finish the research report about polar and nonpolar molecules	❑ *(Optional)* Perform a Hard Water Test
Write	❑ Fill out the experiment sheet	❑ Write about solutions	❑ Color and label the Anatomy of a Water Molecule sketch ❑ Define solution, solubility, solute, and solvent	❑ *(Optional)* Work on memorizing the Law of Constant Composition

Lesson 6 List Schedule

Essentials

Read

- ○ Read Chapter 6: "Addy Walked St. Louis" in *The Route 66 Sticky Note Quest.*
- ○ Choose one of the encyclopedia readings to assign and discuss what was read using the questions provided.

Do

- ○ Do the "Is it polar or nonpolar?" experiment. Fill out the experiment sheet as you do the experiment.

Write

- ○ Write about solutions.
- ○ Look up solution, solubility, solute, and solvent. Add the definitions to the unit vocabulary sheet.
- ○ Label the Anatomy of a Water Molecule sketch.

(Optional) Extras

Read

- ○ Choose another one (or two) of the encyclopedia readings to assign.

Do

- ○ Write a research report on polar and nonpolar molecules.
- ○ Perform a Hard Water Test.

Write

- ○ Work on memorizing the Law of Constant Composition.

Supplies Needed	
Experiments	• 5 Clear cups (or beakers), 5 Plastic spoons, Sugar, Salt, Baking soda, Flour, Petroleum jelly, Water, Vegetable oil, Tablespoon
Optional Activities	• Glass, Tap water, Dish soap

Lesson 6 Science Focus: Solutions

Read: Gathering Information

> **Key Idea**
>
> **A solution includes several different types of molecules evenly mixed.**

Week's Readings

First, spark an interest by having the students read this week's story opener:

- Read Chapter 6 in *The Route 66 Sticky Note Quest*: "Addy Walked St. Louis." (Your student will join Addy and the Whitaker family as they stop in St. Louis, visiting the sites.)

Then follow the story up with some facts. You can choose which resource works best for your students:

- *Usborne Science Encyclopedia*: "Mixtures" pp. 58-59
- *Kingfisher Science Encyclopedia:* "Solutions" pp. 158-159
- *Usborne Illustrated Dictionary of Science:* "Solutions and Solubility" pp. 144-145
- *Everything You Need to Ace Science in One Big Fat Notebook:* "Solutions, Solubility, and Concentration" pp. 85-87

Discussion Questions

After all that reading, use the following questions to guide your discussion time with the students.

1. How do non-polar solvents work? (Non-polar solvents dissolve covalent compounds by pulling the solute molecules away from the lattice, diffusing them between solvent molecules, and holding the solute molecules there using weak Vander Waals forces.)
2. What is the difference between concentrated and dilute solutions? (Concentrated solutions have a large amount of solute in relation to solvent. Dilute solutions have a small amount of solute in relation to solvent.)
3. What is a saturated solution? (A solution is said to be saturated when it cannot hold any more of the solute at a given temperature.)
4. What happens to the solubility of a solid as the temperature increases? (The solubility of a solid increases as the temperature increases.) Is it the same for a gas? (No, the solubility of a gas decreases as the temperature increases.)
5. What is a precipitate? (A precipitate is an insoluble solid that forms in a reaction.)
6. What is a colloid? (A colloid is a mixture of fine particles of different substances that are dispersed evenly and do not dissolve.)

After the discussion time, have the students write a list of facts or a paragraph about what they read. Their writings should include at least a few facts from above.

(Optional) Reading Selections

At another point in the week, the students can check out these optional books:

- *Reactions: An Illustrated Exploration of Elements, Molecules, and Change in the Universe* by Theodore Gray

Do: Experimenting with Science

Experiment: Is it polar or nonpolar?

Materials

You will need the following items:

- ✓ 5 Clear cups (or beakers)
- ✓ 5 Plastic spoons
- ✓ Sugar
- ✓ Salt
- ✓ Baking soda
- ✓ Flour
- ✓ Petroleum jelly
- ✓ Water
- ✓ Vegetable oil
- ✓ Tablespoon

Introduction

Read the following with your students, while they follow along in their lab journal.

> Hazel and Olivia are arguing over solutions when the girls get another sticky note clue. Generally, a solution is made when one type of molecule, which we call the solute, is dissolved in another, which we call the solvent. The molecules can be either polar or nonpolar. Polar molecules have a positive and a negative end, while nonpolar molecules do not carry a charge. As a general rule, like dissolves like. This means that solvents with polar molecules dissolve solutes with polar molecules and vice versa. In this experiment, you are going to use a polar solvent, water, and a nonpolar solvent, vegetable oil, to determine if several substances are polar or nonpolar.

Directions

Have the students do the following experiment to test the polarity of different substances under your supervision.

1. Label the cups #1 through #5.
2. Add ½ cup (120 mL) of room temperature water to each of the five cups.
3. Then stir in 1 TBSP (12.6 g) of sugar to cup #1, 1 TBSP (17.1 g) of salt to cup #2, 1 TBSP (12 g) of baking soda to cup #3, 1 TBSP (5 g) of flour to cup #4, and 1 TBSP (8 g) of petroleum jelly to cup #5.
4. Gently stir each of the cups with a spoon, using a different spoon for each cup. Wait 10 minutes and record your observations and results.
5. Pour out the solutions, rinse well, and dry each of the cups. Add ½ cup (120 mL) of room temperature oil to each of the five cups.
6. Repeat steps 3 and 4.
7. Draw conclusions.

Results and Explanation

When they are done, check the results chart below before reading the results and explanation to the students.

Solute	Solvent: Water	Solvent: Vegetable Oil
Sugar	SS	NS
Salt	S	NS
Baking Soda	S	NS
Flour	NS	SS
Petroleum Jelly	NS	S

***S = Soluble, SS = Semi-soluble, NS = Not soluble**

You should have found that salt, baking soda, and sugar are polar molecules, and that flour and petroleum jelly are both nonpolar molecules.

Discuss how the results from their experiment looked like or didn't look like the expected results. Then, read the following explanation to the students.

Salt and baking soda have ionic bonds, meaning the atoms have exchanged electrons to form a chemical bond that holds the molecule together through the associated charges. These two molecules are very polar substances which can easily dissolve in a polar solvent. Sugar has a long carbon chain that has no associated charge and an OH group that carry a charge. However, it is polar because the pull of the OH group is stronger than the carbon chain, so sugar dissolves better in water than in oil. Flour is composed of thousands of different molecules that come from grinding up grain. Some of these molecules are polar, some are not. On the whole, cooking flour contains more nonpolar molecules, so it distributes more evenly in the oil. Petroleum jelly is a mixture of hydrocarbons, all of which are nonpolar, which is why it dissolves easily in the oil and not in the water.

(Optional) Hands-on Science Activities

At another point in the week, the students can do one or more of the following optional activities.

- **Research Report** – Have the students research more about polar and nonpolar molecules. Then, have them write a one to three paragraph report sharing what they have found.
- **Hard Water Test** – Have the students add a cup (240 mL) of tap water to a glass. Then, add a teaspoon (5 mL) of dish liquid and mix well. Next, use a straw to blow into the water. If lots of bubbles form, then the water is not very hard, but if bubbles do not form, you have hard water.

Write: Keeping a Journal

Journaling Sheets

The students should have already recorded what they learned about solutions on their notes page (LJ p. 57) and completed the experiment sheet (LJ pp. 54-55). On the notes page, the students can also answer the bonus question. Here is an answer for your reference as you discuss this with your students.

- **Bonus Question** – How would make a super-saturated solution that holds more solute than should be possible? (You can make a super-saturated solution by heating the solvent before you add the solute, and then slowly cooling it without disturbing it.)

Vocabulary

At some point in the week, have the students look up the following words and add the definitions to the unit vocabulary sheet on LJ pp. 44-46:

- **Solution** – A homogenous mixture of two or more substances.
- **Solubility** – The ability of a solute to be dissolved.
- **Solute** – The substance that dissolves in the solvent to form a solution.
- **Solvent** – The substance in which the solute dissolves to form a solution, typically a liquid.

Sketch Assignment

Here is the correctly labeled sketch: (LJ p. 56)

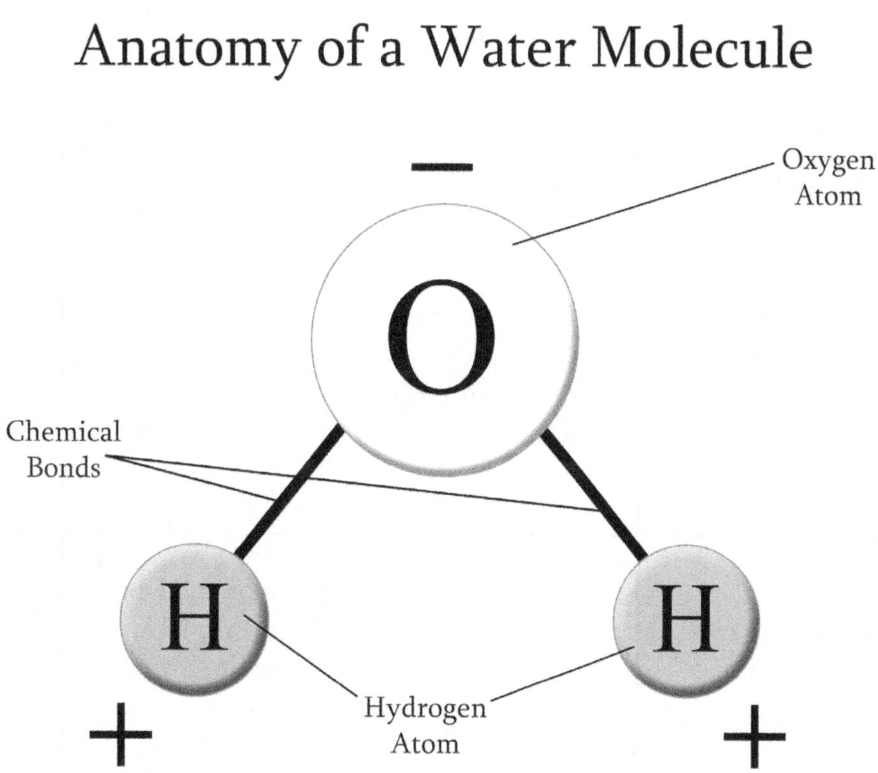

(Optional) Memory Work

At another point in the week, the students can work on the following memory work - The Law of Constant Composition.

Lesson 7 Grid Schedule

Supplies Needed	
Experiments	• 4 Coffee filters, Markers, Alcohol, Coffee can or a wide-mouthed jar, Rubber bands, Eye dropper
Optional Activities	• 4 Cups, 3 Funnels, Water, Dirt, Several different filtering materials (i.e., cotton balls, coffee filter, or gravel) • Water, Salt, Shallow dish

Weekly Schedule

	Day 1	Day 2	Day 3	Day 4
Read	❑ Read Chapter 7: "Addy Traveled Through Missouri" in *The Route 66 Sticky Note Quest*	❑ Choose one of the encyclopedia readings to assign and discuss what was read using the questions provided	❑ *(Optional)* Choose another one of the encyclopedia readings to assign	❑ *(Optional)* Choose another one of the encyclopedia readings to assign
Do	❑ Do the "What colors are in a marker?" experiment	❑ *(Optional)* Do the Muddy Water Filtration activity		❑ *(Optional)* Evaporate Saltwater
Write	❑ Fill out the experiment sheet	❑ Write about separating mixtures	❑ Color and label the Filtration sketch ❑ Define chromatography	❑ *(Optional)* Work on memorizing the Law of Constant Composition

Lesson 7 List Schedule

Essentials

Read

- O Read Chapter 7: "Addy Traveled Through Missouri" in *The Route 66 Sticky Note Quest.*
- O Choose one of the encyclopedia readings to assign and discuss what was read using the questions provided.

Do

- O Do the "What colors are in a marker?" experiment. Fill out the experiment sheet as you do the experiment.

Write

- O Write about separating mixtures.
- O Look up chromatography. Add the definitions to the unit vocabulary sheet.
- O Label the Filtration sketch.

(Optional) Extras

Read

- O Choose another one (or two) of the encyclopedia readings to assign.

Do

- O Do the Muddy Water Filtration activity.
- O Evaporate Saltwater.

Write

- O Work on memorizing the Law of Constant Composition.

Supplies Needed	
Experiments	• 4 Coffee filters, Markers, Alcohol, Coffee can or a wide-mouthed jar, Rubber bands, Eye dropper
Optional Activities	• 4 Cups, 3 Funnels, Water, Dirt, Several different filtering materials (i.e., cotton balls, coffee filter, or gravel) • Water, Salt, Shallow dish

Lesson 7 Science Focus: Separating Mixtures

Read: Gathering Information

> **Key Idea**
>
> **Scientists can separate mixtures in several different ways.**

Week's Readings

First, spark an interest by having the students read this week's story opener:

- Read Chapter 7 in *The Route 66 Sticky Note Quest*: "Addy Traveled Through Missouri." (Your student will follow Addy as the Whitaker family travels from Missouri to Kansas on Route 66, including a stop at a hilarious candy store.)

Then follow the story up with some facts. You can choose which resource works best for your students:

- *Usborne Science Encyclopedia*: "Separating Mixtures" pp. 60-61
- *Kingfisher Science Encyclopedia*: "Separation and Purification" pp. 160-161
- *Usborne Illustrated Dictionary of Science:* "Investigating Substances" pp. 220-221
- *Everything You Need to Ace Science in One Big Fat Notebook:* (no pages scheduled)

Discussion Questions

After all that reading, use the following questions to guide your discussion time with the students.

1. How do decanting and filtration differ? (Decanting allows scientists to separate insoluble solids from a liquid by letting the mixture sit. Filtration separates a liquid from a solid by passing the mixture through a filter that allows the liquid to pass, but not the solids.)
2. Why do scientists use chromatography? (Scientists use chromatography to analyze the substances in a mixture.)
3. What is evaporation? (Evaporation is the process of removing a soluble solid from the solvent by heating the solution.)
4. When is distillation generally used to separate a solution? (Distillation is generally used to separate a mixture when the solvent, the liquid part, of the solution is needed.)
5. How does centrifuging separate a mixture? (Centrifuging separates liquids and solids in a mixture by spinning them around at a high rate of speed.)

After the discussion time, have the students write a list of facts or a paragraph about what they read. Their writings should include at least a few facts from above.

(Optional) Reading Selections

At another point in the week, the students can check out these optional books:

- *Reactions: An Illustrated Exploration of Elements, Molecules, and Change in the Universe* by Theodore Gray

Do: Experimenting with Science

Experiment: What colors are in a marker?

Materials

You will need the following items:

- ✓ 4 Coffee filters
- ✓ Markers
- ✓ Alcohol
- ✓ Coffee can or a wide-mouthed jar
- ✓ Rubber bands
- ✓ Eye dropper

Introduction

Read the following with your students, while they follow along in their lab journal.

> Addy and her sisters use chromatography to make sun catchers. Scientists can use several different methods to separate a mixture into its components. One of those ways is called chromatography, which uses a solvent to move the molecules of a mixture through a medium, such as filter paper. The smaller the molecule in a mixture, the farther a solvent will carry it. So, chromatography is able to separate the molecules in a mixture according to their size. In this experiment, you are going to use chromatography to separate the different colors found in a marker.

Directions

Have the students do the following experiment to observe chromatography in action under your supervision.

1. Choose four different colors of marker to test.
2. In the center of one of the coffee filters, color the outline of a circle with a ½ inch (1.2 cm) diameter with one of the markers you chose.
3. Repeat for the remaining markers.
4. Place one of the colored coffee filters over the top of the coffee can with the circle outline in the center of the opening.
5. Use a rubber band to attach the coffee filter to the can.
6. Then, use the eye dropper to drop 7 to 10 drops of rubbing alcohol in the center of the circle outline on the coffee filter.
7. Wait, observe what happens, and record the colors that are left on the filter paper when the alcohol reaches the edge of the can. (**Note**—If the alcohol does not reach the edge of the can, you may need to add a drop or two more.)
8. Repeat steps 3 and 4 for the remaining colored coffee filters.
9. Record your observations and results before drawing conclusions.

Results and Explanation

When they are done, check the results using the chart below and read the explanation to the students.

Marker Color	Colors Seen	Marker Color	Colors Seen
red	red	**blue**	blue
orange	red, yellow	**purple**	red, blue
yellow	yellow	**black**	red, yellow, blue
green	yellow, blue		

Discuss how the results from their experiment looked like or didn't look like the expected results. Then, read the following explanation to the students.

> Markers can be made up of several different types of ink. Some of those ink molecules are heavier than the others. The alcohol picks up the molecules of ink and carries them along the filter paper. It separates the ink into different colors by depositing the heavier molecules sooner than the lighter ones.

(Optional) Hands-on Science Activities

At another point in the week, the students can do one or more of the following optional activities.

- **Muddy Water Filtration** – Have the students make a solution of water and dirt from outside. Then have them line three funnels with several different filtering materials, such as cotton balls, a coffee filter, and gravel. Place each of the funnels into a glass cup and have them pour a cup of the dirt solution they made. Observe each of the cups to see which filtration material worked the best.
- **Saltwater Evaporation** – Have the students make a solution of ½ cup (120 mL) hot water and 1½ TBSP (25.7 g) of salt. Pour the solution onto a plate and set it in on a window sill in the sun. Check on the plate every 30 minutes. The students should see the water evaporate, leaving salt crystals on the plate.

Write: Keeping a Journal

Journaling Sheets

The students should have already recorded what they learned about the different methods for separating mixtures on their notes page (LJ p. 63) and completed the experiment sheet (LJ pp. 60-61). On the notes page, the students can also answer the bonus question. Here is an answer for your reference as you discuss this with your students.

- **Bonus Question** – If you had a mixture of sand, salt, and iron filings, what would you do to separate each component? (The students answers will vary, but here is one example: You could use a magnet to remove the iron filings, a filter to remove the sand, and evaporation to remove the salt.)

Vocabulary

At some point in the week, have the students look up the following words and add the definitions to the unit vocabulary sheet on LJ pp. 44-46:

- **Chromatography** – A method of separating the substances in a mixture by the rate they move through or along a medium, such as filter paper.

Sketch

Here is the correctly labeled sketch. (LJ p. 62)

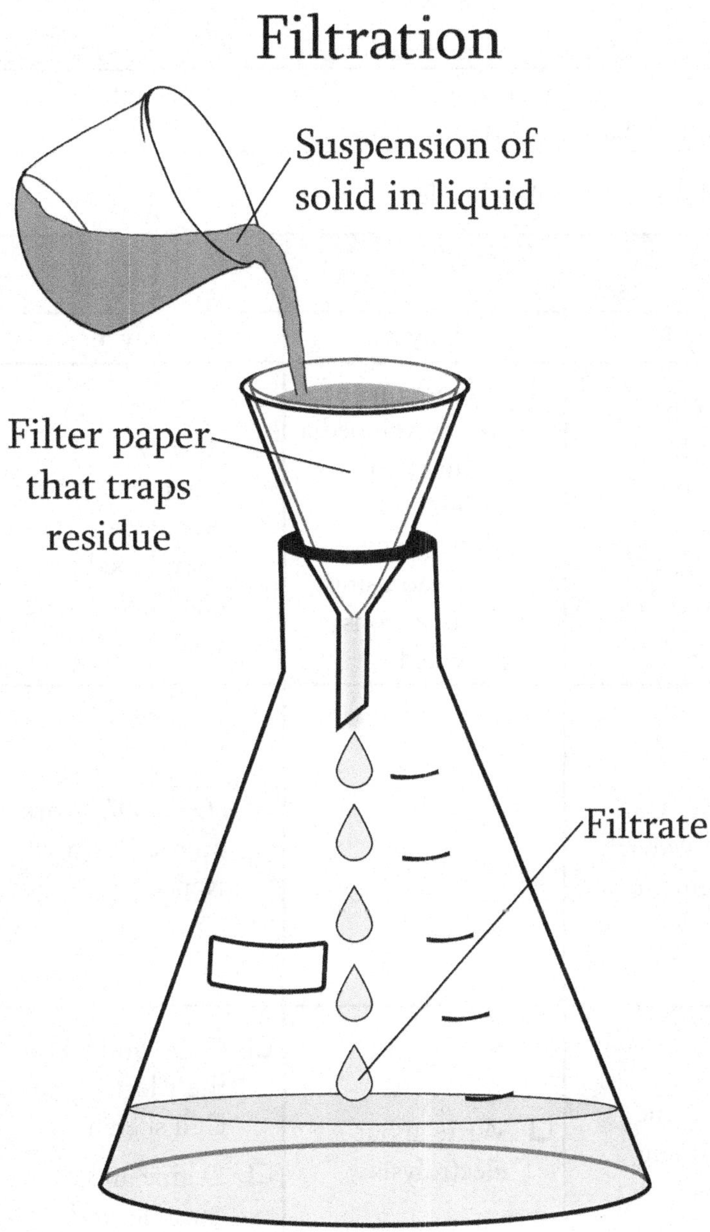

(Optional) Memory Work

At another point in the week, the students can work on the following memory work - The Law of Constant Composition.

Lesson 8 Grid Schedule

Supplies Needed	
Experiments	• Distilled water, 2 Test tubes, Salt, Glass cup, 2 Alligator clips, Covered copper wire, 6v Lantern battery, Permanent marker
Optional Activities	• No additional supplies needed.

Weekly Schedule

	Day 1	Day 2	Day 3	Day 4
Read	❑ Read Chapter 8: "Addy Entered Oklahoma" in *The Route 66 Sticky Note Quest*	❑ Choose one of the encyclopedia readings to assign and discuss what was read using the questions provided	❑ *(Optional)* Choose another one of the encyclopedia readings to assign	❑ *(Optional)* Choose another one of the encyclopedia readings to assign
Do	❑ Do the "Can I split water?" experiment		❑ *(Optional)* Work on the Research Report	❑ *(Optional)* Test Concentration (Experiment Repeat)
Write	❑ Fill out the experiment sheet	❑ Write about electrolysis	❑ Color and label the Electrolytic Cell sketch ❑ Define anode, cathode, and electrolyte	❑ *(Optional)* Work on memorizing the Law of Conservation of Mass

Lesson 8 List Schedule

Essentials

Read

- O Read Chapter 8: "Addy Entered Oklahoma" in *The Route 66 Sticky Note Quest.*
- O Choose one of the encyclopedia readings to assign and discuss what was read using the questions provided.

Do

- O Do the "Can I split water?" experiment. Fill out the experiment sheet as you do the experiment.

Write

- O Write about electrolysis.
- O Look up anode, cathode, and electrolyte. Add the definitions to the unit vocabulary sheet.
- O Label the Electrolytic Cell sketch.

(Optional) Extras

Read

- O Choose another one (or two) of the encyclopedia readings to assign.

Do

- O Work on the Research Report.
- O Test concentration by repeating the experiment.

Write

- O Work on memorizing the Law of Conservation of Mass.

Supplies Needed	
Experiments	• Distilled water, 2 Test tubes, Salt, Glass cup, 2 Alligator clips, Covered copper wire, 6v Lantern battery, Permanent marker
Optional Activities	• No additional supplies needed.

Lesson 8 Science Focus: Electrolysis

Read: Gathering Information

Key Idea

Electrolysis is a way of splitting atoms in a compound with electrical current.

Week's Readings

First, spark an interest by having the students read this week's story opener:

- Read Chapter 8 in *The Route 66 Sticky Note Quest*: "Addy Entered Oklahoma." (Your student will follow Addy and her family as they move through Kansas into Oklahoma, including a stop at the famous Blue Whale!)

Then follow the story up with some facts. You can choose which resource works best for your students:

- *Usborne Science Encyclopedia*: "Electrolysis" pp. 82-83
- *Kingfisher Science Encyclopedia*: (no pages scheduled)
- *Usborne Illustrated Dictionary of Science:* "Electrolysis" pp. 156-157
- *Everything You Need to Ace Science in One Big Fat Notebook:* (no pages scheduled)

Discussion Questions

After all that reading, use the following questions to guide your discussion time with the students.

1. What is an electrolyte, and why do these substances conduct electricity? (An electrolyte is a compound that conducts electricity in a solution or when it is molten. This is because the ions from the compound are free to move. Cations in an electrolyte carry a positive charge, while anions in an electrolyte carry a negative charge.)
2. What is an electrode? (An electrode is a piece of metal or graphite that allows the electrical current to flow into the electrolyte.)
3. What is electrorefining? (Electrorefining is the process of using electrolysis to purify a metal such as copper.)
4. How does electroplating use electrolysis? (Electroplating uses electrolysis to coat a given object with a thin layer of a metal.)
5. How is electrolysis used in extraction? (Electrolysis can be used to extract very reactive metals from impure forms.)

After the discussion time, have the students write a list of facts or a paragraph about what they read. Their writings should include at least a few facts from above.

(Optional) Reading Selections

At another point in the week, the students can check out these optional books:

- *Reactions: An Illustrated Exploration of Elements, Molecules, and Change in the Universe* by Theodore Gray
- *Laws of Electrolysis! What is Electrolysis and More* by Pfiffikus

Do: Experimenting with Science

Experiment: Can I split water?

Materials

You will need the following items:

- ✓ Distilled water
- ✓ 2 Test tubes
- ✓ Salt
- ✓ Glass cup
- ✓ 2 Alligator clips
- ✓ Covered copper wire
- ✓ 6v Lantern battery
- ✓ Permanent marker

CAUTION-

THIS EXPERIMENT SHOULD NOT BE DONE WITHOUT ADULT SUPERVISION!

This experiment has the potential to form hydrogen gas, which is extremely flammable.

Introduction

Read the following with your students, while they follow along in their lab journal.

> The Whitaker family reminisces about the old gas stations along Route 66 over plates of barbecue, which leads to a discussion about electrolysis. Electrolysis separates the components of a molecule within a solution. It uses electrical current to break the compound into its pieces. This process requires that the substance can conduct electricity and that it is either in solution or in a molten state. Electricity passes from two electrodes into the solution, which breaks the compound into its parts. In this experiment, we are going to see if we can use electrolysis to split water.

Directions

Have the students attemp to split water with electrolysis under your supervision. **Caution – Do NOT allow the students to do this experiment on their own as it has two main hazards, electricity with water and potentially flammable hydrogen gas.** It can be done safely in your home as long as you follow the two rules below:

- **Do NOT touch the water when the wire is connected to the battery.** You can wear rubber gloves when you are dealing with the electrical part of this experiment for an extra measure of safety.
- **Do NOT do this experiment anywhere near an open flame.** Also, do not smoke in the house while performing this experiment.

1. Mix together 4 cups (1000 mL) of distilled water with ½ cup (136.5 g) of salt and set the solution aside.
2. Next, cut your wire into two lengths and expose the two ends of each of the wires. Attach one end of the first length of wire to one of the alligator clips. Repeat for the second wire.
3. Now, tape your test tubes to opposite sides of your cup. Fill the cup half of the way full with

your salt water solution. Some of the water will go into your test tubes, which is what you want. They should be fairly close to full of water.

4. Fit the other exposed end of your wire into the bottom of your test tubes, making sure that it does not touch the sides of your test tube. Rest the remaining part of the wire on the side of your container.
5. Then, mark the water level in each of the tubes with a permanent marker and check your set up to make any necessary adjustments.
6. Now, you can attach the alligator clips to each of the battery terminals of the lantern battery. **CAUTION—Do NOT touch the water when the wire is attached to the battery!**
7. Wait for 30 minutes and then detach the wires from the battery. Mark the water level on each of the tubes once again. Remove the tubes and measure the distance between each of your lines to determine how much gas you collected.
8. Record your observations and results before drawing conclusions.

Results and Explanation

When they are done, read the results and explanation to the students.

> You should have seen gas collected in each test tube. You should have also seen twice as much gas in the cathode test tube, which is connected to the positive terminal on their battery

Discuss how the results from their experiment looked like or didn't look like the expected results. Then, read the following explanation to the students.

> We added salt to the water so that it would conduct electricity and allow the process of electrolysis to occur. The bubbles you saw at the tip of the wires in the solution were beads of oxygen and hydrogen gas. These two elements make up the compound water. The oxygen gas forms at the anode, which is the wire connected to the negative terminal on your battery. The hydrogen gas forms at the cathode, which is the wire connected to the positive terminal on your battery. You were able to collect twice as much hydrogen gas than oxygen gas because there are twice as many hydrogen atoms in water than there are oxygen atoms.

(Optional) Hands-on Science Activities

At another point in the week, the students can do one or more of the following optional activities.

- **Testing Concentration (Experiment Repeat)** – You will need the sample set up from the experiment. Test to see if the concentration of salt in the water produces more or less of the two gases. Make sure that you use distilled water to make up your salt water solutions. (The students should see that the more salt they added to the water, the quicker the electrolysis occurred.)
- **Research Report** – Have the students research more about how electrolysis is used in industry. Then, have them write a one to three paragraph report sharing what they have found.

Write: Keeping a Journal

Journaling Sheets

The students should have already recorded what they learned about electrolysis on their notes page (LJ p. 69) and completed the experiment sheet (LJ pp. 66-67). There is no bonus question for this week.

Vocabulary

At some point in the week, have the students look up the following words and add the definitions to the unit vocabulary sheet on LJ pp. 44-46:

- **Anode** – The positively charged electrode by which current enters the cell.
- **Cathode** – The negatively charged electrode by which current leaves the cell.
- **Electrolyte** – A substance that conducts electricity when it is in solution.

Sketch

Here is the correctly labeled sketch: (LJ p. 68)

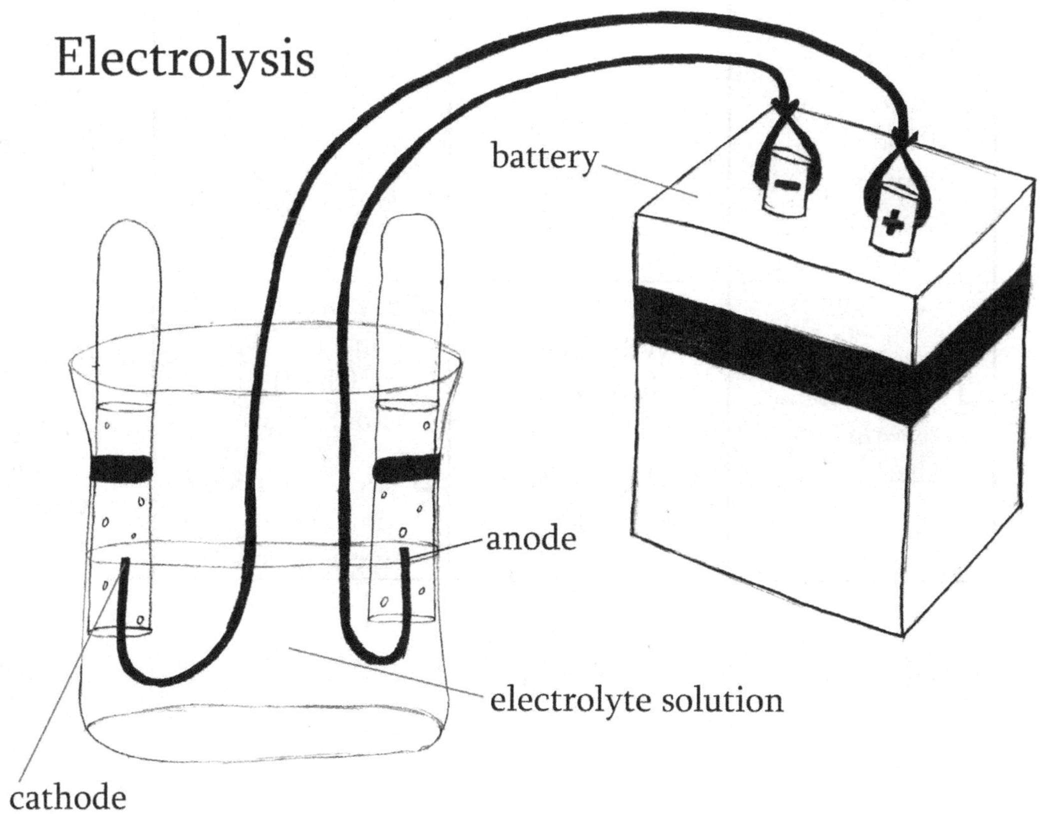

(Optional) Memory Work

At another point in the week, the students can work on the following memory work - The Law of Conservation of Mass.

Lesson 9 Grid Schedule

Supplies Needed				
Experiments	• Cake frosting, Red and yellow bite-sized candies (such as regular sized M&M's)			
Optional Activities	• No additional supplies needed.			
Weekly Schedule				
	Day 1	**Day 2**	**Day 3**	**Day 4**
Read	❑ Read Chapter 9: "Addy Toured the Cowboy Museum" in *The Route 66 Sticky Note Quest*	❑ Choose one of the encyclopedia readings to assign and discuss what was read using the questions provided	❑ *(Optional)* Choose another one of the encyclopedia readings to assign	❑ *(Optional)* Choose another one of the encyclopedia readings to assign
Do	❑ Do the "Chocolate Bonding" activity			❑ *(Optional)* Do the Nomenclature Worksheet
Write	❑ Fill out the experiment sheet	❑ Write about chemical bonding ❑ Define covalent bonding	❑ Color and label the Types of Chemical Bonding sketch ❑ Define ion, ionic bonding, and metallic bonding	❑ *(Optional)* Work on memorizing the Law of Conservation of Mass

Lesson 9 List Schedule

Essentials

Read

- O Read Chapter 9: "Addy Toured the Cowboy Museum" in *The Route 66 Sticky Note Quest.*
- O Choose one of the encyclopedia readings to assign and discuss what was read using the questions provided.

Do

- O Do the "Chocolate Bonding" activity. Fill out the experiment sheet as you do the experiment.

Write

- O Write about chemical bonding.
- O Look up covalent bonding, ion, ionic bonding, and metallic bonding. Add the definitions to the unit vocabulary sheet.
- O Label the Types of Chemical Bonding sketch.

(Optional) Extras

Read

- O Choose another one (or two) of the encyclopedia readings to assign.

Do

- O Do the Nomenclature Worksheet.

Write

- O Work on memorizing the Law of Conservation of Mass.

Supplies Needed	
Experiments	• Cake frosting, Red and yellow bite-sized candies (such as regular sized M&M's)
Optional Activities	• No additional supplies needed.

Lesson 9 Science Focus: Bonding

Read: Gathering Information

Key Idea

Chemical bonding involves the losing or sharing of electrons between two atoms.

Week's Readings

First, spark an interest by having the students read this week's story opener:

- Read Chapter 9 in *The Route 66 Sticky Note Quest*: "Addy Toured the Cowboy Museum." (Your student will follow Addy and the Whitaker family as they explore the Cowboy Museum in Oklahoma City and find another sticky note clue.)

Then follow the story up with some facts. You can choose which resource works best for your students:

- *Usborne Science Encyclopedia*: "Bonding" pp. 68-71
- *Kingfisher Science Encyclopedia*: "Bonding and Valency" pp. 166-167
- *Usborne Illustrated Dictionary of Science:* "Bonding" p. 130, "Ionic Bonding" p. 131, "Covalent Bonding p. 132, "Metallic Bonding" p. 134
- *Everything You Need to Ace Science in One Big Fat Notebook:* "Natural Elements and Ions" pp. 76-77

Discussion Questions

After all that reading, use the following questions to guide your discussion time with the students.

1. What does all bonding involve? (All bonding involves the losing or sharing of electrons from the outermost shells of an atom.)
2. What is a valency electron? (A valency electron is an electron in the outer shell of an atom that is used in the forming of a bond.)
3. What is a cation? (A cation is an ion that is formed when an atom loses an electron in a reaction, resulting in a positive charge.) An anion? (An anion is an ion that is formed when an atom gains an electron in a reaction, resulting in a negative charge.)
4. What happens when an ionic bond forms? (When an ionic bond happens, one atom loses electrons, while the other gains them, which forms oppositely charged ions that are attracted to each other. This attraction causes the atoms to stay together.)
5. How is a covalent bond formed? (A covalent bond is formed when two atoms share electrons between each other so that each atom has a stable outer shell. One pair of shared electrons makes a single covalent bond.)
6. What is metallic bonding? (Metallic bonding occurs when the electrons from the outer shell of a metal atom float around in a "sea" of electrons from other metal atoms.)
7. What is a hydrogen bond? (A hydrogen bond occurs between polar molecules that have a hydrogen and a lone pair of electrons in another molecule.)

After the discussion time, have the students write a list of facts or a paragraph about what they read. Their writings should include at least a few facts from above.

(Optional) Reading Selections

At another point in the week, the students can check out these optional books:

- *Reactions: An Illustrated Exploration of Elements, Molecules, and Change in the Universe* by Theodore Gray
- *Organic Chemistry for Kids: Introducing Chemical Bonding, Hydrocarbons, Functional Groups, Phenyl Ring, Heterocycles to Children* by Jolpic Kidz

Do: Experimenting with Science

Activity: Chocolate Bonding

Materials

You will need the following items:

- ✓ Cake frosting
- ✓ Red and yellow bite-sized candies (such as regular sized M&M's)

Introduction

Read the following with your students, while they follow along in their lab journal.

> Addy finds another sticky note clue with some cryptic words about bonding. There are three main ways that atoms can chemically bond to form molecules. The first is called ionic bonding. In this type of bonding, one atom donates an electron to another to form the bond. The second way is called covalent bonding. In this type of bonding, one atom shares an electron with another atom to form the bond. In this experiment, you are going to use candy to demonstrate both these types of chemical bonding. The third type of bonding is metallic bonding, which we will discuss later.

Directions

Have the students do the following experiment to recreate several types of bonds under your supervision.

Ionic Bonding

1. Count out nine red candies and one yellow candy.
2. Arrange the nine red candies in a circle, so that there are two in the center with seven surrounding them. (*Be sure to leave a space for one more in the outer circle.*) These are the electrons in the outer shell of a chlorine atom.
3. Take the one yellow candy, which represents the electron in the outer shell of a sodium atom, and move into the space in the outer circle of the chlorine electrons. Now you have an ionic bond between your sodium and chlorine atom.

Covalent Bonding

4. Count out eight red candies and two yellow candies.
5. Use the eight red candies to make a model showing the electrons in the outer shell of an oxygen atom by making a circle with six candies on the outside and two on the inside. (*Be sure to leave*

space for two more candies in the outer layer.)

6. Now take two of the yellow candies, which represent the electrons in the outer shells of the two different hydrogen atoms, and bite or cut them in half. Do the same with the two of the red candies from the outer circle you created.
7. Then, use the frosting to glue together one half of each color to form four whole candies.
8. Place the dual colored candies in the outer circle of your model to represent the covalent bond that is found between oxygen and hydrogen in water.
9. Draw a pictures of both the models or take pictures of them.

Results and Explanation

This experiment is meant to be a demonstration of how chemical bonding works so that the concepts will be solidified in your student's mind. The picture below will give you an idea of what their creations should look like:

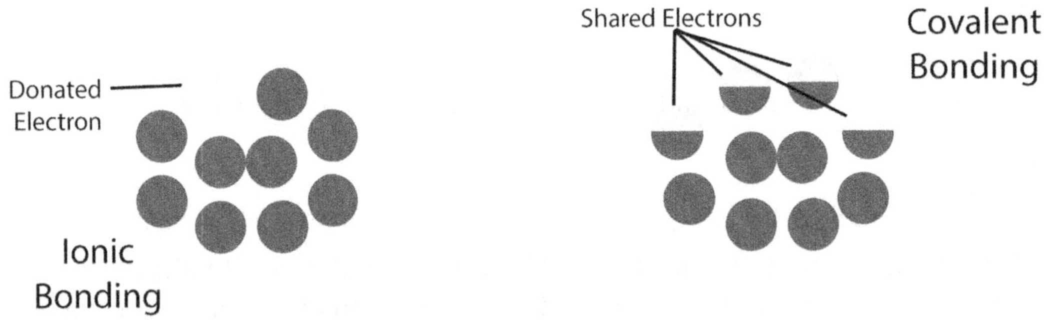

(Optional) Hands-on Science Activities

At another point in the week, the students can do one or more of the following optional activities.

- **Nomenclature –** Nomenclature, or naming compounds in chemistry, came about when scientists wanted a consistent way to refer to the compounds they were discovering. Use the worksheet on pp. 135-136 of the Appendix to teach nomenclature to your student. It contains a very basic explanation of this concept; your student will learn quite a bit more about naming compounds in high school. For now, your goal is to give them a basic understanding of how scientists name chemical compounds.

  Answers to worksheet practice:
  - ✓ **Ionic Compounds:** MgO - magnesium oxide, NaBr - sodium bromide, Li_2S - lithium sulfide, $MgSO_4$ - magnesium sulfate, $Be(OH)_2$ - beryllium hydroxide, $Sr(NO_3)_2$ - strontium nitrate
  - ✓ **Covalent Compounds:** CO_2 - carbon dioxide, NO_2 - nitrogen dioxide, SO_3 - sulfur trioxide, N_2S - dinitrogen sulfide, BF_3 - boron trifluoride, P_2Br_4 - diphosphorus tetrabromide

Write: Keeping a Journal

Journaling Sheets

The students should have already recorded what they learned about the different types of bonding on their notes page (LJ p. 75) and completed the experiment sheet (LJ pp. 72-73). On the notes page, the students can also answer the bonus question. Here is an answer for your reference as you discuss this with your students.

- **Bonus Question** – Does forming a bond require energy or release energy? (Initially, energy is required to break the bonds of the reactants. But forming a chemical bond releases energy. This is because it moves atoms from a high-potential energy state to a more stable, lower-potential energy state.)

Vocabulary

At some point in the week, have the students look up the following words and add the definitions to the unit vocabulary sheet on LJ pp. 44-46:

- **Covalent Bonding** – A chemical bond between two atoms, in which each atom shares an electron.
- **Ion** – An atom or group of atoms that has become charged by gaining or losing one or more electrons.
- **Ionic Bonding** – A strong chemical bond that is formed by the attraction between two ions of opposite charges.
- **Metallic Bonding** – A chemical bond where positive metal ions form a lattice structure with freely moving electrons between them.

Sketch

Here is the correctly labeled sketch. (LJ p. 74)

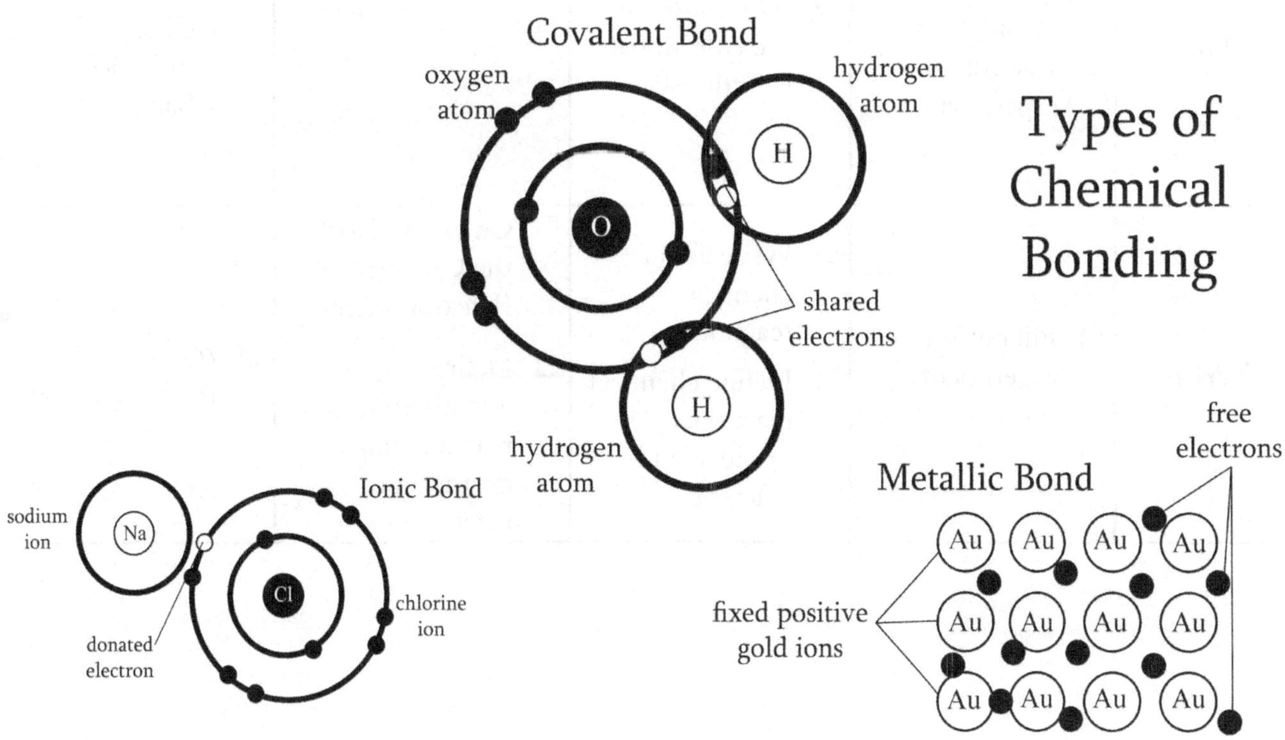

(Optional) Memory Work

At another point in the week, the students can work on the following memory work - The Law of Constant Composition.

Lesson 10 Grid Schedule

Supplies Needed	
Experiments	• Yeast, Hydrogen peroxide, Epsom salts, Water, 2 Cups, 2 Thermometers
Optional Activities	• Laundry detergent, Baking soda, Water • Posterboard, Markers or pictures

Weekly Schedule

	Day 1	Day 2	Day 3	Day 4
Read	❑ Read Chapter 10: "Addy Solved One Clue" in *The Route 66 Sticky Note Quest*	❑ Choose one of the encyclopedia readings to assign and discuss what was read using the questions provided	❑ *(Optional)* Choose another one of the encyclopedia readings to assign	❑ *(Optional)* Choose another one of the encyclopedia readings to assign
Do	❑ Do the "Exothermic or Endothermic?" experiment	❑ *(Optional)* Do the Additional Reactions		❑ *(Optional)* Make a Chemical vs. Physical Changes poster
Write	❑ Fill out the experiment sheet	❑ Write about chemical reactions ❑ Define chemical reaction, product, and reactant	❑ Color and label the Chemical Reaction sketch ❑ Define endothermic reaction and exothermic reaction	❑ *(Optional)* Take the Unit 2 Test

Lesson 10 List Schedule

Essentials

Read

- O Read Chapter 10: "Addy Solved One Clue" in *The Route 66 Sticky Note Quest.*
- O Choose one of the encyclopedia readings to assign and discuss what was read using the questions provided.

Do

- O Do the "Exothermic or Endothermic?" experiment. Fill out the experiment sheet as you do the experiment.

Write

- O Write about chemical reactions.
- O Look up chemical reaction, endothermic reaction, exothermic reaction, product, and reactant. Add the definitions to the unit vocabulary sheet.
- O Label the Chemical Reaction sketch.

(Optional) Extras

Read

- O Choose another one (or two) of the encyclopedia readings to assign.

Do

- O Do the Additional Reactions.
- O Make a Chemical vs. Physical Changes poster.

Write

- O Take the Unit 2 Test.

Supplies Needed	
Experiments	• Yeast, Hydrogen peroxide, Epsom salts, Water, 2 Cups, 2 Thermometers
Optional Activities	• Laundry detergent, Baking soda, Water • Posterboard, Markers or pictures

Lesson 10 Science Focus: Reactions

Read: Gathering Information

Key Idea

In a chemical reaction, bonds are broken and made.

Week's Readings

First, spark an interest by having the students read this week's story opener:

- Read Chapter 10 in *The Route 66 Sticky Note Quest*: "Addy Solved One Clue." (Your student will follow Addy as the Whitaker family travels from Oklahoma to Texas on Route 66. They also finally find out what the blank sticky note says.)

Then follow the story up with some facts. You can choose which resource works best for your students:

- *Usborne Science Encyclopedia*: "Chemical Reactions" pp. 76-77
- *Kingfisher Science Encyclopedia*: "Chemical Reactions" pp. 162-163
- *Usborne Illustrated Dictionary of Science:* "Energy and Chemical Reactions" pp. 162-163
- *Everything You Need to Ace Science in One Big Fat Notebook:* "Physical and Chemical Properties and Changes" pp. 62-65

Discussion Questions

After all that reading, use the following questions to guide your discussion time with the students.

1. What are the substances in a reaction called? (The substances present at the beginning of the reaction are the reactants. The substances present at the end of the reaction are the products.)
2. What happens in a chemical reaction, and what does it require? (In a chemical reaction, bonds are broken and made. This process requires energy.)
3. What is the difference between exothermic and endothermic reactions? (In an endothermic reaction, the heat energy needed to break the bonds of the reactants is greater than the heat energy given off when the new bonds of the products are formed. In an exothermic reaction, the heat energy needed to break the bonds of the reactants is less than the heat energy given off when the new bonds of the products are formed.)
4. What is activation energy? (Activation energy is the amount of energy a reaction needs to start.)
5. What is the law of conservation of mass? (The law of conservation of mass says that matter cannot be created or destroyed during a chemical reaction.)
6. How do chemists show reactions? (Chemists show reactions with equations that have the reactants on the right and the products on the left.)
7. How do chemists measure chemical substances? (Chemists measure chemical substances using moles.)

After the discussion time, have the students write a list of facts or a paragraph about what they read. Their writings should include at least a few facts from above.

(Optional) Reading Selections

At another point in the week, the students can check out these optional books:

- 📖 *Reactions: An Illustrated Exploration of Elements, Molecules, and Change in the Universe* by Theodore Gray
- 📖 *Chemical Reactions Part of: Science Readers* by Jenna Winterberg

Do: Experimenting with Science

Experiment: Exothermic or Endothermic?

Materials

You will need the following items:

- ✓ Yeast
- ✓ Hydrogen peroxide
- ✓ Epsom salts
- ✓ Water
- ✓ 2 Cups
- ✓ 2 Thermometers

Introduction

Read the following with your students, while they follow along in their lab journal.

> Addy and her sisters reveal the message on the blank sticky note with a chemical reaction. In a chemical reaction, bonds are broken and reformed. Every chemical reaction involves energy that is either released or absorbed in the form of heat, light, or electricity. We call the heat-releasing reactions exothermic reactions and the heat-absorbing reactions endothermic reactions. In this experiment, you are going to look at two different reactions to determine if they are exothermic or endothermic.

Directions

Have the students do the following experiment to observe chromatography in action under your supervision.

1. Add 1 cup (240 mL) of hydrogen peroxide to one of the cups and 1 cup (240 mL) of warm water to the other.
2. Place a thermometer in each cup and wait 5 minutes.
3. Record the temperature of both cups as the starting temperature.
4. Now, add 1 TBSP (8.5 g) of yeast to the hydrogen peroxide and gently stir.
5. Then, add 1 TBSP (14.2 g) of Epsom salts to the water and gently stir.
6. Wait until the temperature on the thermometer stabilizes and record the final temperature for the reaction.
7. Draw conclusions.

Results and Explanation

When they are done, read the results and explanation to the students.

> You should have seen the yeast and hydrogen peroxide reaction was exothermic and the Epsom salts and water reaction was endothermic.

Discuss how the results from their experiment looked like or didn't look like the expected results. Then, read the following explanation to the students.

> Hydrogen peroxide is a very powerful oxidizing agent and it immediately begins to breakdown the cell walls of the yeast. This process releases energy into the solution in the form of heat. On the other hand, it requires energy to break the bonds of the magnesium sulfate in the Epsom salts. The energy is absorbed by the magnesium and sulfate ions and removed from the water, which cause the temperature to drop.

(Optional) Hands-on Science Activities

At another point in the week, the students can do one or more of the following optional activities.

- **Additional Reactions** – Have the students mix laundry detergent and water or baking soda and water to see if they are exothermic or endothermic reactions. (Both of these reactions are exothermic.)
- **Chemical vs. Physical Changes Poster** – Have the students think of some changes that they see every day, such as water being boiled and making a cup of tea. Have them list these changes and then decide if they are chemical or physical changes. (Note—Remember that a chemical change looks different and it is usually a permanent irreversible change. A physical change can look different, but chemically it is still the same. It is not permanent and can be reversed.) Once they have at least five of each, have them create a poster depicting the difference between chemical and physical changes.

Write: Keeping a Journal

Journaling Sheets

The students should have already recorded what they learned about chemical reactions on their notes page (LJ p. 81) and completed the experiment sheet (LJ pp. 78-79). On the notes page, the students can also answer the bonus question. Here is an answer for your reference as you discuss this with your students.

- **Bonus Question** – When milk turns sour, how can we prove this is a chemical change? (Souring is an irreversible reaction where lactose is converted into lactic acid, creating new substances. This makes it a chemical change, not a physical change.)

Vocabulary

At some point in the week, have the students look up the following words and add the definitions to the unit vocabulary sheet on LJ pp. 44-46:

- **Chemical Reaction** – An interaction between the atoms of two substances where the atoms rearrange to form two new molecules.
- **Endothermic Reaction** – A chemical reaction that takes in heat.
- **Exothermic Reaction** – A chemical reaction that gives off heat.
- **Product** – A new substance produced in a chemical reaction.
- **Reactant** – The substance present at the beginning of a chemical reaction.

Sketch

Here is the correctly labeled sketch. (LJ p. 80)

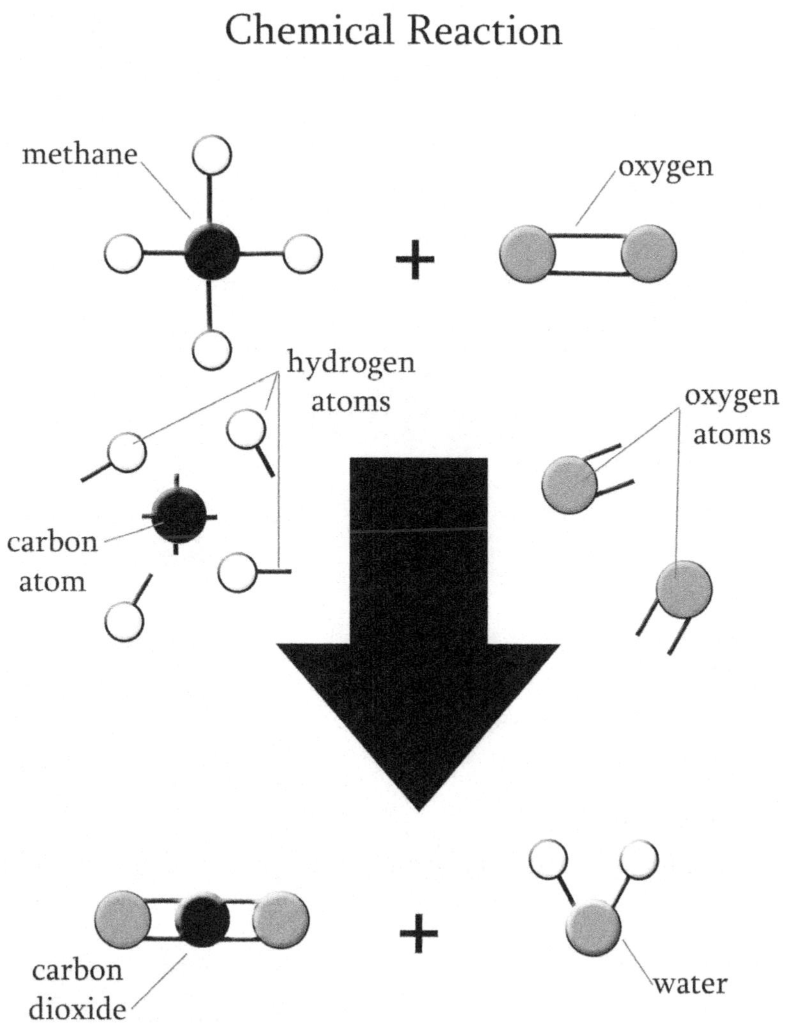

(Optional) Solutions and Reactions Test

The students can complete the optional unit test found on pp. 156-159 in the Appendix. The answers for this test are found on pp. 80-81. A printable copy of the test can be downloaded for free from:

- https://elementalscience.com/blogs/resources/sq200

Solutions and Reactions Test Answers

Vocabulary Matching (2 points each)

1. T
2. J
3. K
4. R
5. Q
6. H
7. P
8. N
9. L
10. F
11. E
12. I
13. A
14. D
15. G
16. U
17. B
18. M
19. O
20. C
21. S

Note -
If you need a digital copy of the unit test, you can download one for free from this program's resource page. Just head to:
https://elementalscience.com/blogs/resources/sq200

True and False (2 points each)

1. False (A mixture is composed of elements or compounds that have not been chemically joined.)
2. True
3. True
4. True
5. False (Evaporation is a way to separate a soluble solid that has been dissolved in a liquid solvent.)
6. True
7. True
8. True

9. True
10. False (Chemical bonding always involves the outermost electrons of an atom.)
11. False (The substances present at the beginning of the reaction are the reactants. The substances present at the end of the reaction are the products.)
12. True

Short Answer

1. (2 points) The law of constant composition states that a pure compound always contains the same elements in the same proportions.
2. (2 points) Concentrated solutions have a large amount of solute in relation to solvent. Dilute solutions have a small amount of solute in relation to solvent.
3. (6 points) Students' answers should include two of the following: Evaporation is a way to separate a soluble solid that has been dissolved in a liquid solvent. Filtration is used to separate a suspension of particles of a solid that are spread through a liquid. Centrifuging is used to separate a suspension when filtering or settling will not work. Distillation is used to separate a mixture of two or more liquids. Fractional distillation is used to separate two or more liquids when the liquids have similar boiling points.
4. (4 points) An electrolyte is a compound that conducts electricity in a solution or when it is molten. This is because the ions from the compound are free to move. Cations in an electrolyte carry a positive charge, while anions in an electrolyte carry a negative charge.
5. (6 points) Ionic bonding is a strong chemical bond that is caused by the attraction between two ions of opposite charges. Covalent bonding is a chemical bond between two atoms, in which each atom shares an electron. Metallic bonding is a chemical bond where positive metal ions create a lattice structure with freely moving electrons between them.
6. (4 points) In an endothermic reaction the heat energy needed to break the bonds of the reactants is greater than the heat energy given off when the new bonds of the products are formed. In an exothermic reaction the heat energy needed to break the bonds of the reactants is less than the heat energy given off when the new bonds of the products are formed.

Total points: 90

SciQuest

Addy's Adventure Teacher Notes

Scientist Biography Unit

What you'll study in this unit-

- 11: The Scientist of Your Choice (p. 84)

Memory Work-

None

Lesson 11 Scientist Biography

Read: Gathering Information

Week's Readings

First, spark an interest by having the students read this week's story opener:

- Read Chapter 11 in *The Route 66 Sticky Note Quest*: "Addy Reached Halfway on the Mother Road." (Your student will follow Addy and the Whitaker family as they cross into New Mexico on Route 66, passing the halfway point on the Mother Road.)

> Note–
> There is no schedule for this lesson as it depends on the length of the book your students' choose. We suggest you work with your students to spread the work across the week.

Step 1: Choose A Scientist

After the stduents read Addy's chapter, have them choose a scientist to study for the week. You can have the student choose a book from your local library or do a pre-prepared scientist biography unit for this week. We recommend choosing one of the Science Chunks units we offer, all of which also include hands-on science activites related to the scientist:

- https://elementalscience.com/collections/scientist-biography-units

The following information is from the student's lab journal:

> During the next few weeks, you are going to be researching and learning more about a scientist that has contributed to science. You need to begin your scientist biography project by choosing which scientist you will research. You can choose someone from the field of chemistry, like Marie Curie, or one of the lesser known chemists, such as Frederick Wohler, or someone that has interested you.

Step 2: Research the Scientist

The following information is from the student's lab journal:

> Once you have chosen the scientist you would like to study, you can begin your research. Begin by looking for a biography on your chosen scientist at the library. Then, look for articles on the chemist in magazines, newspapers, encyclopedias, or on the Internet. You will need to know the following about your scientist to write your report:
>
> - Biographical information on the scientist (where he or she was born, his or her parents, siblings, and how he or she grew up)
> - The scientist's education (where he or she went to school, what kind of student he or she was, what he or she studied, and so on)
> - His or her scientific contributions (the research that he or she

participated in, any significant discoveries he or she made, and the state of the world at the time of his or her contributions)

As you read over the material you have gathered, be sure to write down any facts you glean in your own words. You can do this on a separate sheet or on separate index cards - just make sure you have at least five pieces of information for each category. And don't forget to record a list of the resources you have used for your research.

Step 3: Create an Outline

The following information is from the student's lab journal:

Now that your research is completed, you are ready to begin the process of writing a report on your chosen scientist. You are going to organize the notes you took during step two into a formal outline which you will use next week to write the rough draft of your report. Use the outline template provided below as a guide. You should include information on why you chose the particular scientist in your introduction section. For the conclusion section of the outline, you need to include why you believe someone else should learn about your chosen scientist and your impression of the scientist (i.e. Did you like the scientist? Do you feel that they made a significant impact on their field of science?).

Step 4: Write a Rough Draft

The following information is from the student's lab journal:

Finally you are ready to take your outline and turn it into a rough draft. Simply take the points on your outline and combine or add in sentence openers and transitions to create a cohesive paragraph. Here's what your rough draft should look like:

- Paragraph 1 (from outline point I): introduce the scientist;
- Paragraph 2 (from outline point II): tell about the scientist's education;
- Paragraph 3-5 (from outline point III): share the scientist's contributions (one paragraph for each contribution);
- Paragraph 6 (from outline point IV): share your thoughts on the scientist and why someone should learn about him or her.

You can choose to hand write or type up your rough draft on a separate sheet of paper. However, keep in mind that you will need a typed version for step five.

Step 5: Revise to Create a Final Draft

The following information is from the student's lab journal:

Now that you have a typed, double-spaced rough draft, look over it one more time to make any changes you would like. Then, have your teacher or one of your peers look over the paper for you to correct any errors and bring clarity to any of the unclear sections. Once this is complete, make the necessary changes to your paper to create your final draft. Print out your final paper to share or use it to create a book or poster on the scientist.

You can have the students do an oral presention of their scientists if time allows.

SciQuest

Addy's Adventure Teacher Notes

Forces and Motion Unit

What you'll study in this unit-

Memory Work-

Newton's 3 Laws of Motion (LJ p. 93)

Lesson 12 Grid Schedule

Supplies Needed	
Experiments	• Thick, sturdy cardboard, 1 Brad fastener, Rubber band, Hole punch or nail, String – about 3 in (10 cm), 3 Jumbo paper clips, Pen, Objects of varying weight
Optional Activities	• A rope (or access to a computer with Internet)

Weekly Schedule				
	Day 1	**Day 2**	**Day 3**	**Day 4**
Read	❑ Read Chapter 12: "Addy Explored Albuquerque" in *The Route 66 Sticky Note Quest*	❑ Choose one of the encyclopedia readings to assign and discuss what was read using the questions provided	❑ *(Optional)* Choose another one of the encyclopedia readings to assign	❑ *(Optional)* Read one or both of the optional reading selections
Do	❑ Do the "Can I measure force?" experiment	❑ *(Optional)* Do the Resultant Forces Worksheet		❑ *(Optional)* Play Tug of War
Write	❑ Fill out the experiment sheet	❑ Write about forces	❑ Color and label the Force Diagram sketch ❑ Define balance, force, force field, and newton	❑ *(Optional)* Work on memorizing Newton's Three Laws of Motion

Lesson 12 List Schedule

Essentials

Read

- O Read Chapter 12: "Addy Explored Albuquerque" in *The Route 66 Sticky Note Quest.*
- O Choose one (or more) of the encyclopedia readings to assign and discuss what was read using the questions provided.

Do

- O Do the "Can I measure force?" experiment. Fill out the experiment sheet as you do the experiment.

Write

- O Write about forces.
- O Look up balance, force, force field, and newton. Add the definitions to the unit vocabulary sheet.
- O Label the Force Diagram sketch.

(Optional) Extras

Read

- O Read one or both of the optional reading selections.

Do

- O Do the Resultant Forces Worksheet.
- O Play Tug of War.

Write

- O Work on memorizing Newton's Three Laws of Motion

Supplies Needed	
Experiments	• Thick, sturdy cardboard, 1 Brad fastener, Rubber band, Hole punch or nail, String – about 3 in (10 cm), 3 Jumbo paper clips, Pen, Objects of varying weight
Optional Activities	• A rope (or access to a computer with Internet)

Lesson 12 Science Focus: Forces

Read: Gathering Information

> **Key Idea**
>
> **A force is a push or pull that acts on an object.**

Week's Readings

First, spark an interest by having the students read this week's story opener:

- Read Chapter 12 in *The Route 66 Sticky Note Quest*: "Addy Explored Albuquerque." (Your student will watch as the Whitaker family explores parts of New Mexico, finding another note.)

Then follow the story up with some facts. You can choose which resource works best for your students:

- *Usborne Science Encyclopedia*: "Force" pp. 118-121
- *Kingfisher Science Encyclopedia:* "Force" pp. 290-291
- *Usborne Illustrated Dictionary of Science:* "Force" pp. 6-7
- *Everything You Need to Ace Science in One Big Fat Notebook:* "Force and Net Force" pp. 99-100

Discussion Questions

After the reading, use the following questions to guide your discussion time with the students.

1. What does a force do? (A force influences the shape and/or motion of an object.)
2. What is a vector quantity? (A vector quantity has both magnitude and direction.)
3. What are the four main types of forces? Describe each. (The main types of forces are gravitational, magnetic, electric, and nuclear. Gravitational force is the attraction between two objects that have mass. Magnetic force is the force between two moving charges. Electric force is the force between two electrically charged particles. Nuclear force is the force of attraction between the particles of an atomic nucleus.)
4. What are intermolecular forces? (Intermolecular forces are the electromagnetic forces between two molecules.)
5. What is a frictional force? (Frictional force is a force that acts to oppose the motion of two touching objects or surfaces.) (**Note**—The students will be covering frictional forces in greater detail in lesson 14.)

After the discussion time, have the students write a list of facts or a paragraph about what they read. Their writings should include at least a few facts from above.

(Optional) Reading Selections

At another point in the week, the students can check out these optional books:

- *A Crash Course in Forces and Motion with Max Axiom, Super Scientist* by Emily Sohn, Steve Erwin, and Charles Barnett III (This book is also scheduled for lesson 16 as an optional reading selection.)
- *Balanced and Unbalanced Forces* by Jenna Winterberg
- *Forces at the Amusement Park (Amusement Park Science)* by Tammy Enz

Do: Experimenting with Science

Experiment: Can I measure force?

Note-
Be sure that the students use thick, sturdy cardboard when making their force meter or it can tear. And DO NOT throw away the force meter. You will need it for the experiment in lesson 14.

Materials

You will need the following items:

- ✓ Thick, sturdy cardboard
- ✓ 1 Brad fastener
- ✓ Rubber band
- ✓ Hole punch or nail
- ✓ String – about 3 in (10 cm)
- ✓ 3 Jumbo paper clips
- ✓ Pen
- ✓ Objects of varying weight

Introduction

Read the following with your students, while they follow along in their lab journal.

Addy and her sisters have to use force to discover their next clue. Forces are all around us. They push and pull objects, causing them to move or change shape. In today's experiment, you are going to create your own force meter that can measure the amount of force an object exerts. In a force meter, an object applies a downward force, which stretches a rubber band or spring. We can measure the amount of displacement to determine how much force was applied.

Directions

Have the students do the following experiment to make a meter to measure force under your supervision.

Troubleshooting Tips-
To see a visual representation of a homemade force meter and how to calibrate it, check out the following video:
https://www.youtube.com/watch?v=jwCwwKLa0GE

1. Assemble the force meter.
   - Cut out a 3.5 in (9 cm) by 12 in (31 cm) rectangle from the cardboard.
   - Then, punch a hole with the hole punch or nail near the top, large enough for the brad fastener to slide through.
   - Slip one of the paper clips through the brad, through the hole, and fasten the brad on the opposite side.
   - Slide the rubber band onto the opposite end of the paper clip.
   - Next, take another paper clip and turn out a portion of the end to make a pointer.
   - Tie the string to one end of the pointer paper clip and then slide the other end onto the rubber band.
   - Take the third paper clip and fashion a hook out of it.

- Once you are done, attach the hook to the other end of the string.
- Hold your force meter at the top and mark where the pointer rests. This line will be your zero force mark.
- Now draw a scale down the remainder of your force meter. You can use finger widths, inches, or centimeters for your scale, just as long as you use the same measurement for each mark. (**Note**—You will need your force meter for next week's experiment as well.)

2. Now that the force meter is assembled, you can use it to measure the different objects. Simply attach each object to the hook and observe what happens.
3. Write down how much the rubber band stretched on the chart below.
4. Repeat this process for each of your objects.

Results and Explanation

When they are done, read the results and explanation to the students

> You should have seen that the heavier object will cause the rubber band to stretch farther than a lighter object.

Discuss how the results from their experiment looked like or didn't look like the expected results. Then, read the following explanation to the students.

> The weight of each object acts as a force that pulls down on the rubber band, causing it to stretch. The more the object weighs, the greater the force, which causes the rubber band to stretch farther.

(Optional) Hands-on Science Activities

At another point in the week, the students can do one or more of the following optional activities.

- **Tug of War** – In a tug of war, each team is using force to pull the other team across the line. One team's pulling force cancels out the other team's pulling force, which keeps the players at a stand-still. That is, until one team's pulling force is greater than the other's! This week, explain to your students how force plays a role in tug of war and then let them try it out for themselves. If you can't get a team together, have the students do the tug of war simulation from the PhET website.

  https://phet.colorado.edu/en/simulations/forces-and-motion

- **Resultant Force Worksheet** – Have the students complete the resultant force worksheet on Appendix pp. 137-138.

  Answers

  1. Resultant force = 0N, object is in balance
  2. Resultant force = -2N, object will begin moving in the opposite direction
  3. Resultant force = 8N, object will continue in the same direction
  4. Resultant force = 0N, object is in balance

WRITE: KEEPING A JOURNAL

JOURNALING SHEETS

The students should have already recorded what they learned about forces on their notes page (LJ p. 103) and completed the experiment sheet (LJ pp. 100-101). On the notes page, the students can also answer the bonus question. Here is an answer for your reference as you discuss this with your students.

- **Bonus Question** – What forces affect a parachute as it drifts down to the ground? A boat as it floats on water? (A parachute is affected by gravity and air resistance as it drifts toward the ground. A boat is affected by gravity and buoyancy as it floats on water. Students may also add air or water resistance to the list of forces affecting the boat.)

Vocabulary

At some point in the week, have the students look up the following words and add the definitions to the unit vocabulary sheet on LJ pp. 94-97:

- **Balance** – A state of equilibrium when the forces acting on an object cancel each other out; also known as a zero resultant force.
- **Force** – A push or pull that acts on an object.
- **Force field** – The area in which a force can be felt.
- **Newton** – The measurement of force; one newton is the force is takes to move a one kilogram object at one meter per second squared (1 N = 1 kg • 1 m/s2).

Sketch Assignment

Here is the correctly labeled sketch: (LJ p. 102)

Force Diagram

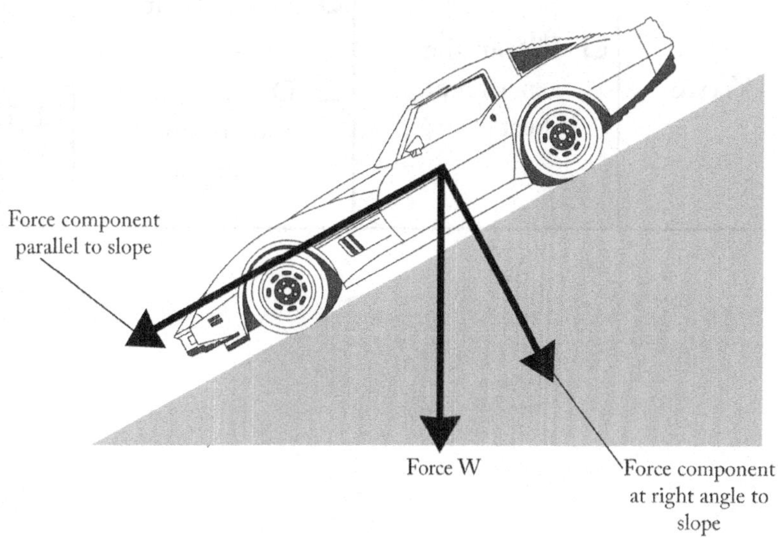

(OPTIONAL) MEMORY WORK

At another point in the week, the students can work on the following memory work - Newton's Three Laws of Motion.

- **First Law** – An object will not move unless a force, like a push or pull, moves it. Once it is moving, an object will not stop moving in a straight line unless it is forced to change.
- **Second Law** – A force acting on an object will change its motion. The greater the force on an object, the greater the change in its motion.
- **Third Law** – For every reaction, there is an equal but opposite reaction.

Lesson 13 Grid Schedule

Supplies Needed	
Experiments	• Goldfish cracker, Small marshmallow, Piece of lettuce, Piece of bacon fat, Aluminum pan, Matches, Safety glasses, Bucket of water
Optional Activities	• No additional supplies needed.

Weekly Schedule

	Day 1	Day 2	Day 3	Day 4
Read	❑ Read Chapter 13: "Addy Zoomed into Arizona" in *The Route 66 Sticky Note Quest*	❑ Choose one of the encyclopedia readings to assign and discuss what was read using the questions provided	❑ *(Optional)* Choose another one of the encyclopedia readings to assign	❑ *(Optional)* Choose another one of the encyclopedia readings to assign
Do	❑ Do the "Do different types of food contain different amounts of energy?" experiment			❑ *(Optional)* Do the Work Worksheet
Write	❑ Fill out the experiment sheet	❑ Write about energy ❑ Define energy and energy conversion	❑ Color and label the Energy Chain sketch ❑ Define potential energy, kinetic energy, and work	❑ *(Optional)* Work on memorizing Newton's Three Laws of Motion

Lesson 13 List Schedule

Essentials

Read

- O Read Chapter 13: "Addy Zoomed into Arizona" in *The Route 66 Sticky Note Quest.*
- O Choose one of the encyclopedia readings to assign and discuss what was read using the questions provided.

Do

- O Do the "Do different types of food contain different amounts of energy?" experiment. Fill out the experiment sheet as you do the experiment.

Write

- O Write about energy.
- O Look up energy, energy conversion, potential energy, kinetic energy, and work. Add the definitions to the unit vocabulary sheet.
- O Label the Energy Chain sketch.

(Optional) Extras

Read

- O Choose another one (or two) of the encyclopedia readings to assign.

Do

- O Do the Work Worksheet.

Write

- O Work on memorizing Newton's Three Laws of Motion.

Supplies Needed	
Experiments	• Goldfish cracker, Small marshmallow, Piece of lettuce, Piece of bacon fat, Aluminum pan, Matches, Safety glasses, Bucket of water
Optional Activities	• No additional supplies needed.

Lesson 13 Science Focus: Energy

Read: Gathering Information

> **Key Idea**
>
> **Energy is the ability to do work.**

Week's Readings

First, spark an interest by having the students read this week's story opener:

- Read Chapter 13 in *The Route 66 Sticky Note Quest*: "Addy Zoomed into Arizona." (Your student will follow Addy and the Whitaker family as they roll into Arizona, visiting the Petrified National Forest.)

Then follow the story up with some facts. You can choose which resource works best for your students:

- *Usborne Science Encyclopedia*: "Energy" pp. 106-107
- *Kingfisher Science Encyclopedia:* "Potential and Kinetic Energy" pp. 292-293
- *Usborne Illustrated Dictionary of Science:* "Energy" pp. 8-9
- *Everything You Need to Ace Science in One Big Fat Notebook:* "Energy" pp. 130-134

Discussion Questions

After all that reading, use the following questions to guide your discussion time with the students.

1. How are work and energy related? (Energy is the ability to do work, and work cannot be done without energy.)
2. Can energy be destroyed? (No, according to the law of conservation of energy, energy cannot be created or destroyed.) If not, what happens to it? (Energy changes forms, which is known as energy conversion or transformation.)
3. What is the SI unit of measurement for energy? (The SI unit for energy is the Joule.)
4. What are some common forms of energy? (Several common forms of energy include chemical, heat, wave, radiation, nuclear, and electrical.) Explain a bit more about each. (Chemical energy is energy stored in food, fuels, and batteries. Heat is energy that flows from one place to another due to a change in temperature. Wave energy is the energy associated with wave action. Radiation is energy in the form of electromagnetic waves or streams of particles. Nuclear energy is the energy stored in the nucleus of an atom. Electrical energy is the energy associated with moving electric charge.)
5. What is the difference between kinetic and potential energy? (Kinetic energy is the energy of motion, but potential energy is the energy that an object has because of its position.)
6. What is power? (Power is the rate of doing work, or rather the rate of change of energy.)

After the discussion time, have the students write a list of facts or a paragraph about what they read. Their writings should include at least a few facts from above.

(Optional) Reading Selections

At another point in the week, the students can check out these optional books:

- *Energy (See for Yourself)* by Chris Woodford

Do: Experimenting with Science

Experiment: Do different types of food contain different amounts of energy?

Materials

You will need the following items:

- ✓ Goldfish cracker
- ✓ Small marshmallow
- ✓ Piece of lettuce
- ✓ Piece of bacon fat
- ✓ Aluminum pan
- ✓ Matches
- ✓ Safety glasses
- ✓ Bucket of water

CAUTION-

THIS EXPERIMENT SHOULD NOT BE DONE WITHOUT ADULT SUPERVISION!

This experiment uses matches, which should not be used without adult supervision.

Introduction

Read the following with your students, while they follow along in their lab journal.

> The Whitaker family visited a mine for ore used to make nuclear energy. As we break down the food we eat, it releases energy in the form of calories. Some of the chemicals in food are broken down quickly, releasing a burst of energy. Other chemicals take a bit more work to break down, so they release energy slowly over time. These reactions work together to fuel our bodies' processes over time. In today's experiment, you will be testing different types of food to see how much energy they release.

Directions

Have the students do the following experiment to test the energy of different types of food under your supervision.

1. Head outside and set the aluminum pan on a concrete or tile surface or inside a grill.
2. Place the cracker in the aluminum pan and have an adult use the match to light the cracker.
3. Time how long it takes for the cracker to burn and write this on the chart.
4. Use the bucket of water to extinguish any smoldering embers. Then, clean and dry the aluminum pan.
5. Now, repeat the procedure from step 1 through 4 for the marshmallow, lettuce, and bacon fat. (**Note**—Make sure that your samples are similar in size, so that your results are more accurate.)
6. Draw conclusions.

Results and Explanation

When they are done, read the results and explanation to the students

> You should have seen the marshmallow burned the quickest, followed by the lettuce, the goldfish cracker, and finally the bacon.

Discuss how the results from their experiment looked like or didn't look like the expected results. Then, read the following explanation to the students.

> The marshmallow has a quick sugar-rush of energy, which burns and then quickly dies out. The lettuce has fiber and a bit of sugar, which burns slower than the marshmallow. However, since lettuce is mostly water, it doesn't burn all that long. The goldfish cracker has a mixture of sugar, starch, and fiber, which keeps it burning longer than the two previous samples. The bacon fat is full of lipids and proteins that burn for a long time. Just like the food we eat, vegetables give us the fuel our bodies needs, but we have to eat a lot of bulk to survive on only vegetables. Sweets, like marshmallows, give our bodies a quick punch of energy that doesn't last very long. Carbohydrates rich in fiber fuel our bodies for a much longer period of time than sugar. Fats and proteins provide our bodies with a longlasting source of energy. Since our bodies need more than just fuel, it is best to have a balance of fruits and vegetables, carbohydrates, along with fats and proteins.

(Optional) Hands-on Science Activities

At another point in the week, the students can do one or more of the following optional activities.

- **Work Worksheet** – Have the students complete the work worksheet on Appendix pp. 139-140.

  Answers
  1. The man has done 200 J of work.
  2. The person has done 240 J of work.
  3. The hot air has done 630,000 J of work.
  4. The winch has done 1050 J of work.

Write: Keeping a Journal

Journaling Sheets

The students should have already recorded what they learned about energy on their notes page (LJ p. 109) and completed the experiment sheet (LJ pp. 106-107). On the notes page, the students can also answer the bonus question. Here is an answer for your reference as you discuss this with your students.

- **Bonus Question** – What is the difference between renewable and nonrenewable energy sources? (Renewable energy sources will not run out, whereas nonrenewable energy sources will eventually run out.)

Vocabulary

At some point in the week, have the students look up the following words and add the definitions to the unit vocabulary sheet on LJ pp. 94-97:

- **Energy** – The ability to do work.
- **Energy Conversion** – The process of changing one form of energy into another.

- **Kinetic Energy** – The energy of an object in motion; it depends upon the object's mass and speed.
- **Potential Energy** – The energy that an object has stored; it depends upon the object's weight and height.
- **Work** – The transfer of energy that occurs when a force moves or changes an object.

Sketch Assignment

Here is the correctly labeled sketch: (LJ p. 108)

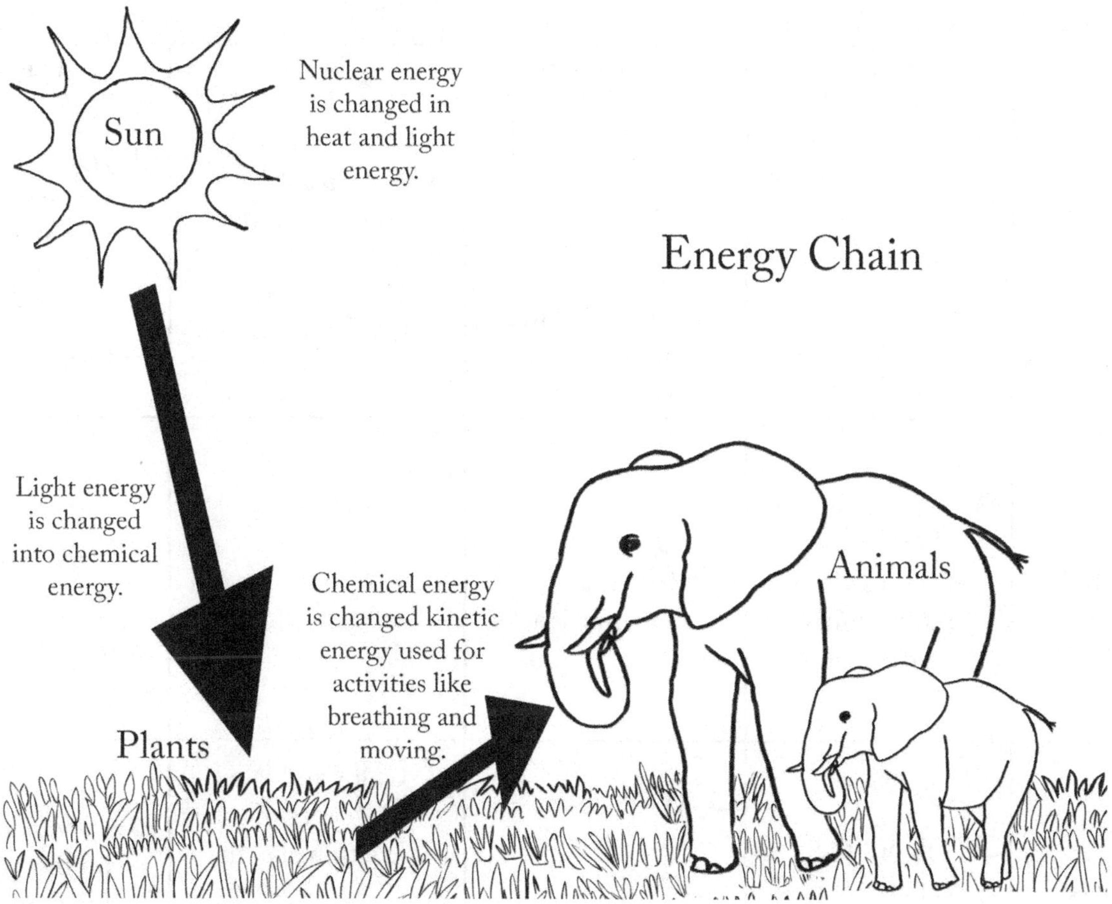

(Optional) Memory Work

At another point in the week, the students can work on the following memory work - Newton's Three Laws of Motion.

Lesson 14 Grid Schedule

Supplies Needed	
Experiments	• Force Meter from Lesson 12, Small wooden block (a.k.a. Jenga block), Eye-hook screw, Sandpaper, Felt, Foil, Spray oil, Tape measure
Optional Activities	• Marble, Marble Track • Several Objects from Nature

Weekly Schedule				
	Day 1	**Day 2**	**Day 3**	**Day 4**
Read	❑ Read Chapter 14: "Addy Took the Train to the Grand Canyon" in *The Route 66 Sticky Note Quest*	❑ Choose one of the encyclopedia readings to assign and discuss what was read using the questions provided	❑ *(Optional)* Choose another one of the encyclopedia readings to assign	❑ *(Optional)* Choose another one of the encyclopedia readings to assign
Do	❑ Do the "How does friction affect movement?" experiment	❑ *(Optional)* Recreate Galileo's Tracks		❑ *(Optional)* Test gravity outdoors
Write	❑ Fill out the experiment sheet	❑ Write about friction and gravity ❑ Define air resistance	❑ Color and label the Types of Friction sketch ❑ Define friction, gravity, and terminal velocity	❑ *(Optional)* Work on memorizing Newton's Three Laws of Motion

Lesson 14 List Schedule

Essentials

Read

- O Read Chapter 14: "Addy Took the Train to the Grand Canyon" in *The Route 66 Sticky Note Quest.*
- O Choose one of the encyclopedia readings to assign and discuss what was read using the questions provided.

Do

- O Do the "How does friction affect movement?" experiment. Fill out the experiment sheet as you do the experiment.

Write

- O Write about friction and gravity.
- O Look up air resistance, friction, gravity, and terminal velocity. Add the definitions to the unit vocabulary sheet.
- O Label the Types of Friction sketch.

(Optional) Extras

Read

- O Choose another one (or two) of the encyclopedia readings to assign.

Do

- O Recreate Galileo's Tracks.
- O Test gravity outdoors.

Write

- O Work on memorizing Newton's Three Laws of Motion.

Supplies Needed	
Experiments	• Force Meter from Lesson 12, Small wooden block (a.k.a. Jenga block), Eye-hook screw, Sandpaper, Felt, Foil, Spray oil, Tape measure
Optional Activities	• Marble, Marble Track • Several Objects from Nature

Lesson 14 Science Focus: Friction and Gravity

Read: Gathering Information

> **Key Idea**
>
> **Friction is a force that opposes the motion of objects as they touch.**

Week's Readings

First, spark an interest by having the students read this week's story opener:

- Read Chapter 14 in *The Route 66 Sticky Note Quest*: "Addy Took the Train to the Grand Canyon." (Your student will follow Addy and the Whitaker family as they visit the Grand Canyon.)

Then follow the story up with some facts. You can choose which resource works best for your students:

- *Usborne Science Encyclopedia*: "Friction" pp. 124-125
- *Kingfisher Science Encyclopedia*: "Relativity and Gravity" pp. 298-299, "Friction" pp. 308-309
- *Usborne Illustrated Dictionary of Science:* "Gravitation" pp. 18-19
- *Everything You Need to Ace Science in One Big Fat Notebook:* , "Gravity and Friction" pp. 109-112

Discussion Questions

After all that reading, use the following questions to guide your discussion time with the students.

1. What two things affect the force of gravity? (The distance between the objects and the mass of the objects both affect the force of gravity.)
2. Is weight constant? Why or why not? (Weight is not constant because it is dependent upon the distance from and mass of the closest planet.)
3. What is escape velocity? (Escape velocity is the minimum velocity an object must travel so that it can escape the gravitational pull of a planet.)
4. What does it mean to be weightless? (When an object does not exert a force on its surroundings, it is said to be weightless.)
5. How does the roughness of a surface affect the amount of friction? (The rougher a surface, the stronger the force of friction.)
6. Is friction useful? (Yes, in some cases, such as braking or lighting a match, friction is useful. In other cases, such as in between machine parts, friction can be damaging.)
7. What does a lubricant do? (A lubricant reduces friction between surfaces.)
8. What is drag? (Drag is the friction that occurs between the air and an object moving through it.)
9. What does a streamlined design do? (A streamlined design reduces drag so that an object can move more easily.)

After the discussion time, have the students write a list of facts or a paragraph about what they read. Their writings should include at least a few facts from above.

(Optional) Reading Selections

At another point in the week, the students can check out these optional books:

- 📖 *Drag! Friction and Resistance (TIME For Kids Informational Text)* by Stephanie Paris
- 📖 *The Gravity Tree: The True Story of a Tree That Inspired the World* by Anna Crowley Redding

Do: Experimenting with Science

Experiment: How does friction affect movement?

Materials

You will need the following items:

- ✓ Force Meter from Lesson 12
- ✓ Small wooden block (a.k.a. Jenga block)
- ✓ Eye-hook screw
- ✓ Sandpaper
- ✓ Felt
- ✓ Foil
- ✓ Spray oil
- ✓ Tape measure

Introduction

Read the following with your students, while they follow along in their lab journal.

> Addy's father briefly mentions gravity as they overlook the rim of the Grand Canyon, but what about friction? Remember that when an object is in forward motion, several forces are acting on it. There is the driving force, which is propelling the object forward. There is weight (or gravity), which pulls the object downward. There is air resistance, which slows the object down. Finally, there is friction. In today's experiment, you are going to act as the driving force for a block as it moves across a track. Then, you are going to use a variety of materials to test how friction affects the motion of the block.

Directions

Have the students do the following experiment to observe the effect of friction under your supervision.

1. Screw the eyehook screw into the top of the wooden block. Then, attach it to the hook on the force meter so that the block can be dragged horizontally.
2. Next, use the tape measure to mark off a 1-foot (0.3 meter) track on a smooth surface, like a table our counter.
3. Now, place the block at the beginning of your track with the force meter in front over the track.
4. Pull the block from the force meter evenly to the end in three seconds.
5. Observe how much the rubber band on the force meter

> **Troubleshooting Tip-**
>
> The following video shows how this experiment should be set up:
>
> 💻 https://m.youtube.com/watch?v=HP8H3HWBrZE

stretched and record that on your experiment sheet.

6. Then, place the piece of sandpaper on your track before repeating steps 3 to 5.
7. Next, remove the sandpaper and cover the track with felt before repeating steps 3 to 5.
8. Finally, remove the felt. Cover the track with foil and spray with oil before repeating steps 3 to 5.
9. Record your observations and results before drawing conclusions.

Results and Explanation

When they are done, read the results and explanation to the students

> You should have seen that more force was needed to pull the block when it was on the felt and sandpaper. And that less force was needed to pull the block when it was on the oil-covered foil.

Discuss how the results from their experiment looked like or didn't look like the expected results. Then, read the following explanation to the students.

> Both the felt and the sandpaper increase the amount of friction that acts on the block as it slides over the track. The oil-coated foil reduces the amount of friction that acts on the block as it slides over the track. Friction is a force that opposes the motion of an object as it passes another. So, when friction increases, the object will slow down. Conversely, when friction is decreased, the object will speed up.

(Optional) Hands-on Science Activities

At another point in the week, the students can do one or more of the following optional activities.

- **Galileo's Tracks** – Have the students study the effect of friction using a marble track, just like Galileo did. Have the students set up a track and send the marble down it several times. Each time, have them record the time it takes to get to the bottom. Then, have the student sprinkle some sand or salt all over the track. Have them send the marbles down several more times, recording the time it takes. (The students should see that the marbles were much slower the second time, due to the amount of friction that was created by the sand or salt on the track.)
- **Gravity** – Have the students test gravity using several objects from nature. Begin by taking a walk outside with the students. Have them look for several objects in nature that are round and about the same size. The objects should have different weights, such as a piece of fruit, a rock, and a nut. Once you get home, have the students hold each of the round objects in their hands and drop them at the same time. What happened? (They should see that both of the objects hit the ground at the very same time. If you can do this safely from a porch or balcony that will give you a bit more height, your results will be even more amazing.)

Write: Keeping a Journal

Journaling Sheets

The students should have already recorded what they learned about friction and gravity on their notes page (LJ p. 115) and completed the experiment sheet (LJ pp. 112-113). On the notes page, the students can also answer the bonus question. Here is an answer for your reference as you discuss this with your students.

- **Bonus Question** – What is the difference between mass and weight? (Mass is the measurement of how much matter an object contains, whereas weight is the measurement of the pull of gravity on an object. The more mass an object contains, the more it weighs because there is more substance on which gravity can pull.)

Vocabulary

At some point in the week, have the students look up the following words and add the definitions to the unit vocabulary sheet on LJ pp. 94-97:

- **Air resistance** – The force that air exerts on an object as it falls.
- **Friction** – A force that opposes the motion of objects that touch as they move past each other.
- **Gravity** – The force that acts between two masses; it is an attractive force.
- **Terminal Velocity** – The point at which the force acting on an object of air resistance is equal to the force of gravity acting on the object.

Sketch

Here is the correctly labeled sketch. (LJ p. 114)

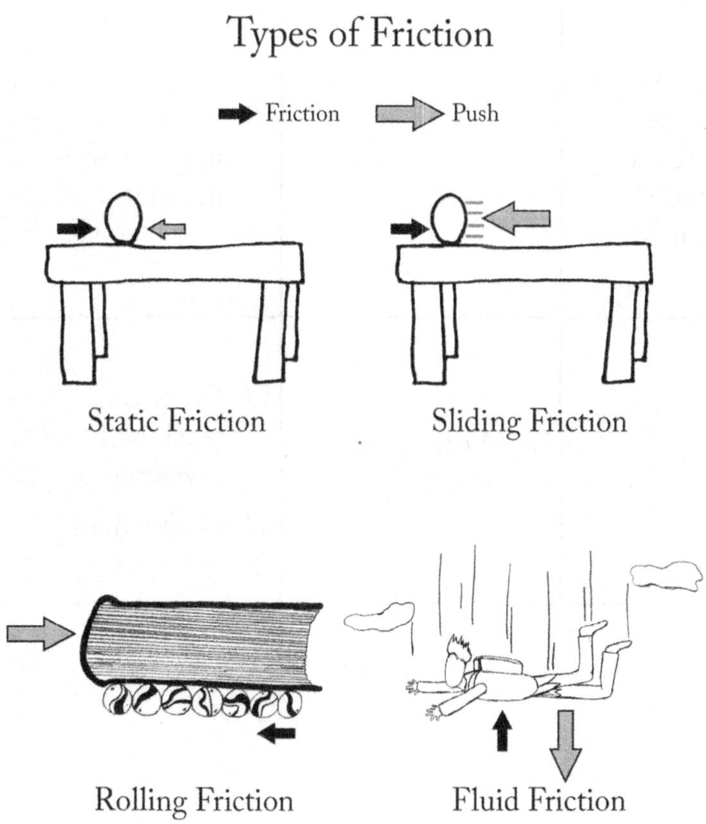

(Optional) Memory Work

At another point in the week, the students can work on the following memory work - Newton's Three Laws of Motion.

Lesson 15 Grid Schedule

Supplies Needed	
Experiments	• 2 Tin cans (one large, one small), Screw, Water, Piece of clay, Cup measurer, Tape measurer
Optional Activities	• 2-Liter Soda bottle, Screw, Water, Piece of clay, Cup measurer, Tape measurer • Posterboard, Markers

Weekly Schedule				
	Day 1	**Day 2**	**Day 3**	**Day 4**
Read	❑ Read Chapter 15: "Addy Noticed the Knee" in *The Route 66 Sticky Note Quest*	❑ Choose one of the encyclopedia readings to assign and discuss what was read using the questions provided	❑ *(Optional)* Choose another one of the encyclopedia readings to assign	❑ *(Optional)* Choose another one of the encyclopedia readings to assign
Do	❑ Do the "What affects the pressure exerted by a fluid?" experiment		❑ *(Optional)* Learn more about forces in fluids	❑ *(Optional)* Do the Large Scale Pressure Test
Write	❑ Fill out the experiment sheet	❑ Write about pressure	❑ Color and label the Fluid Dynamics sketch ❑ Define fluid	❑ *(Optional)* Work on memorizing Newton's Three Laws of Motion

Lesson 15 List Schedule

Essentials

Read

- O Read Chapter 15: "Addy Noticed the Knee" in *The Route 66 Sticky Note Quest.*
- O Choose one of the encyclopedia readings to assign and discuss what was read using the questions provided.

Do

- O Do the "What affects the pressure exerted by a fluid?" experiment. Fill out the experiment sheet as you do the experiment.

Write

- O Write about pressure.
- O Look up fluid. Add the definition to the unit vocabulary sheet.
- O Label the Fluid Dynamics sketch.

(Optional) Extras

Read

- O Choose another one (or two) of the encyclopedia readings to assign.

Do

- O Learn more about forces in fluids.
- O Do the Large Scale Pressure Test.

Write

- O Work on memorizing Newton's Three Laws of Motion.

Supplies Needed	
Experiments	• 2 Tin cans (one large, one small), Screw, Water, Piece of clay, Cup measurer, Tape measurer
Optional Activities	• 2-Liter Soda bottle, Screw, Water, Piece of clay, Cup measurer, Tape measurer • Posterboard, Markers

Lesson 15 Science Focus: Pressure

Read: Gathering Information

> **Key Idea**
>
> **Pressure is the amount of force pushing on a given area.**

Week's Readings

First, spark an interest by having the students read this week's story opener:

- Read Chapter 15 in *The Route 66 Sticky Note Quest*: "Addy Noticed the Knee." (Your student will join Addy and her family as they wait for the Bluebird bus to be repaired, finding a new clue in the sticky note quest.)

Then follow the story up with some facts. You can choose which resource works best for your students:

- *Usborne Science Encyclopedia*: "Pressure" pp. 132-133
- *Kingfisher Science Encyclopedia*: "Fluids" p. 310, "Pressure" p. 311
- *Usborne Illustrated Dictionary of Science:* "Pressure" p. 25
- *Everything You Need to Ace Science in One Big Fat Notebook:* "Pressure" pp. 87-88

Discussion Questions

After all that reading, use the following questions to guide your discussion time with the students.

1. How are pressure, force, and area related? (If you spread force over a large area, you will reduce the amount of pressure exerted. If you concentrate the force over a small area, you will increase the amount of pressure exerted.)
2. How do barometers work? (Barometers measure atmospheric pressure through fluid levels. When the air pressure increases, the fluid level rises, showing the high pressure on the scale.)
3. What is upthrust, and how is it related to weight? (Upthrust is the upward force an object in a fluid experiences. According to the Archimedes' principle, upthrust is equal to the weight of the fluid displaced.)
4. What does the principle of flotation tell us? (The principle of floatation tells us that an object will float if the upthrust is equal to the object's weight.)

After the discussion time, have the students write a list of facts or a paragraph about what they read. Their writings should include at least a few facts from above.

(Optional) Reading Selections

At another point in the week, the students can check out these optional books:

- *Pressure, Heat and Temperature - Physics for Kids* by Baby Professor
- *Pop! Air and Water Pressure* by Stephanie Paris
- *Air And Water Pressure* by Richard Spilsbury

Do: Experimenting with Science

Experiment: What affects the pressure exerted by a fluid?

Materials

You will need the following items:

- ✓ 2 Tin cans (one large, one small)
- ✓ Screw
- ✓ Water
- ✓ Piece of clay
- ✓ Cup measurer
- ✓ Tape measurer

NOTE-

The cans the students use in this test need to have different radiuses. The small can should have a smaller radius than the larger one. In other words, the students should choose a tomato paste can and a soup can, not a tuna can and soup can.

Introduction

Read the following with your students, while they follow along in their lab journal.

> Addy learns about pressure and fluids as she helps her father fix the Bluebird's engine. A fluid, such as a liquid or a gas, fills in the available space within a container. The fluid exerts pressure on the surfaces of the container throughout the space it occupies. The amount of that pressure depends upon several factors. In today's experiment, you will be looking at two factors – area and force – to see how they affect the pressure a fluid exerts in a container.

Directions

Have the students observe pressure in action under your supervision.

1. Fill the small tin can with water and then use a cup measurer to determine how much water the can holds.
2. Use the screw to poke a hole in the bottom of the small tin can about ½ inch (1.25 cm) from the bottom.
3. Put a piece of clay over the hole to prevent any water spilling out before filling the can with water.
4. Lay out the tape measure on a surface that you don't mind getting wet, preferably outdoors.
5. Place the can at the first mark on the tape measure, and quickly remove the clay plug.
6. Measure how far the water goes and record it on the chart.
7. Repeat this process two more times and then average the distances.
8. Repeat steps 1 through 7 using the larger can.
9. Draw conclusions.

Results and Explanation

When they are done, read the results and explanation to the students.

You should have seen that the water in the smaller can went further than the water in the large can.

Discuss how the results from their experiment looked like or didn't look like the expected results. Then, read the following explanation to the students.

We examined the effect of area on pressure. The force exerted by the water remains the same because they used the same amount of water for each test. Area and pressure are inversely proportional – the greater the area, the lower the pressure. This is why the water that shot out of the smaller can with the lesser area went further.

(Optional) Hands-on Science Activities

At another point in the week, the students can do one or more of the following optional activities.

- **Forces in Fluids** – In addition to the Archimedes principle, which students learned about in their reading, there are two other principles about forces in fluids known as Pascal's and Bernoulli's principles. Have the students research and learn about these principles. They can present their findings in a paragraph or on a poster depicting the two principles. (Pascal's principle states that fluids transmit pressure equally in all directions. Bernoulli's principle states that a fast-moving fluid has a lower pressure than a slow-moving one.)
- **Large Scale Pressure Test (Experiment Repeat)** – Use the screw to poke three holes in the 2-Liter bottle – one about 1 inch (2.5 cm) from the bottom, one at the middle of the bottle, and one 2 inches (5 cm) from the top of the bottle. Plug each of the holes with clay and fill the bottle to the very top. Place the full 2-Liter bottle at the first mark on the tape measure. Quickly remove the clay plugs. Measure how far the water goes from each of the holes and record it on the experiment sheet. Repeat this process two more times and then average the distances.

Write: Keeping a Journal

Journaling Sheets

The students should have already recorded what they learned about pressure on their notes page (LJ p. 121) and completed the experiment sheet (LJ pp. 118-119). On the notes page, the students can also answer the bonus question. Here is an answer for your reference as you discuss this with your students.

- **Bonus Question** – Why doesn't a person feel the miles of air above crushing down on them? (The fluid pressure in air doesn't crush us because there is typically just as much pressure inside your body pressing outward as there is air pressure outside your body pushing inward. They typically cancel out, meaning that there is no overall force on you and you don't get crushed.)

Vocabulary

At some point in the week, have the students look up the following word and add the definition to the unit vocabulary sheet on LJ pp. 94-97:

- **Fluid** – A substance that assumes the shape of the container it is in, such as a liquid or gas.

Sketch

Here is the correctly labeled sketch: (LJ p. 120)

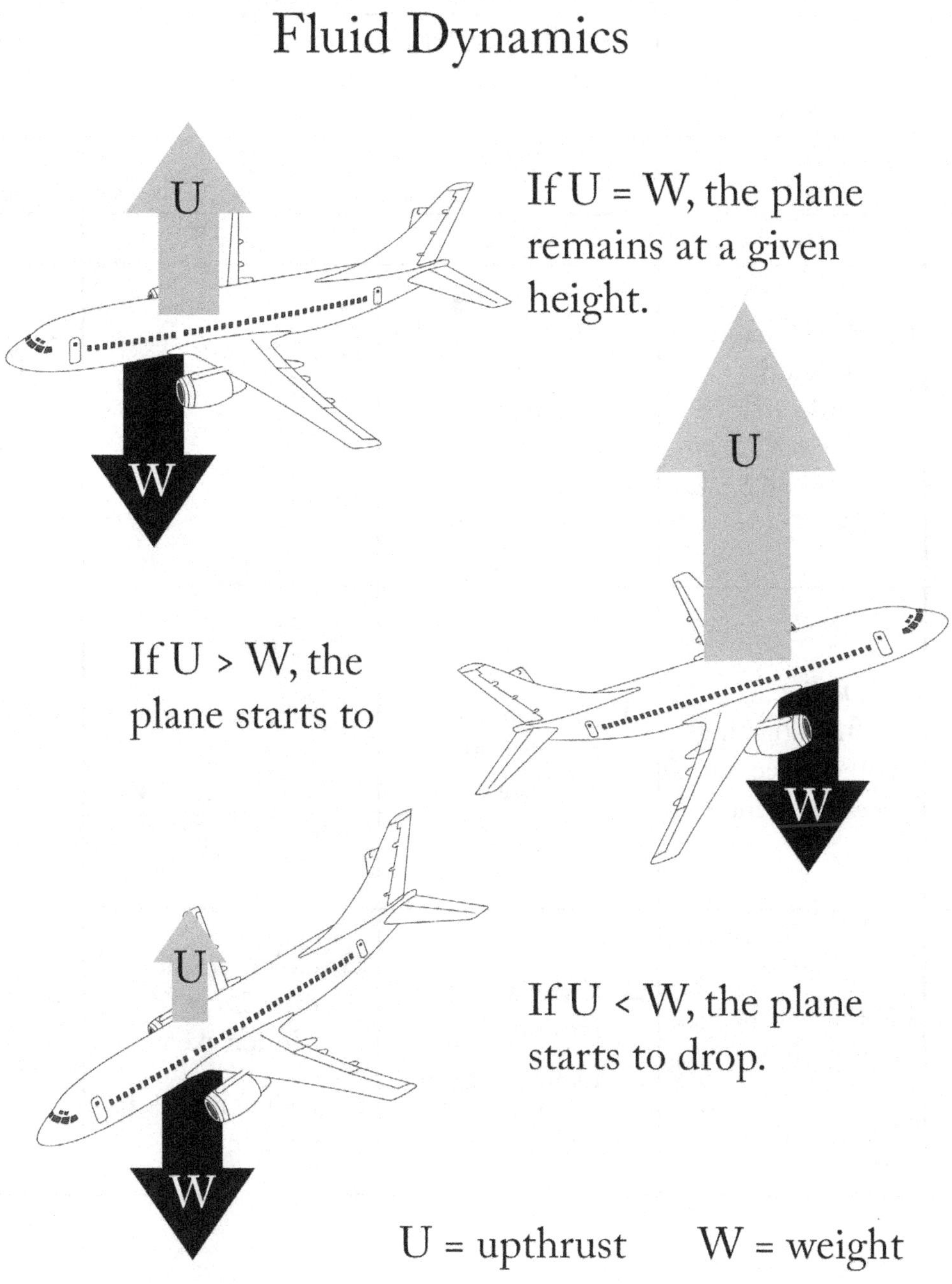

(Optional) Memory Work

At another point in the week, the students can work on the following memory work - Newton's Three Laws of Motion.

Lesson 16 Grid Schedule

Supplies Needed	
Experiments	• Jenga block with the eyehook from last week, String, 2 Toy cars, Egg
Optional Activities	• Egg, Various shock absorbing materials (such as cotton balls, newspaper, packing peanuts or fabric), a 1 quart plastic container (the type that fruit is typically packed in), Masking tape

Weekly Schedule				
	Day 1	**Day 2**	**Day 3**	**Day 4**
Read	❑ Read Chapter 16: "Addy Settled on the Answer" in *The Route 66 Sticky Note Quest*	❑ Choose one of the encyclopedia readings to assign and discuss what was read using the questions provided	❑ *(Optional)* Choose another one of the encyclopedia readings to assign	❑ *(Optional)* Choose another one of the encyclopedia readings to assign
Do	❑ Do the "Investigating the Three Laws" experiment	❑ *(Optional)* Do the Second Law Worksheet		❑ *(Optional)* Make an Egg Drop Carrier
Write	❑ Fill out the experiment sheet	❑ Write about motion ❑ Define mass and weight	❑ Color and label the 3 Laws of Motion sketch ❑ Define inertia and momentum	❑ *(Optional)* Work on memorizing Newton's Three Laws of Motion

LESSON 16 LIST SCHEDULE

ESSENTIALS

Read

- O Read Chapter 16: "Addy Settled on the Answer" in *The Route 66 Sticky Note Quest.*
- O Choose one of the encyclopedia readings to assign and discuss what was read using the questions provided.

Do

- O Do the "Investigating the Three Laws" experiment. Fill out the experiment sheet as you do the experiment.

Write

- O Write about motion.
- O Look up inertia, mass, weight, and momentum. Add the definitions to the unit vocabulary sheet.
- O Label the 3 Laws of Motion sketch.

(OPTIONAL) EXTRAS

Read

- O Choose another one (or two) of the encyclopedia readings to assign.

Do

- O Do the Second Law Worksheet.
- O Make an Egg Drop Carrier.

Write

- O Work on memorizing Newton's Three Laws of Motion.

SUPPLIES NEEDED	
Experiments	• Jenga block with the eyehook from last week, String, 2 Toy cars, Egg
Optional Activities	• Egg, Various shock absorbing materials (such as cotton balls, newspaper, packing peanuts or fabric), a 1 quart plastic container (the type that fruit is typically packed in), Masking tape

Lesson 16 Science Focus: Motion

Read: Gathering Information

Key Idea

Motion is a change in position. Dynamics is the study of motion.

Week's Readings

First, spark an interest by having the students read this week's story opener:

- Read Chapter 16 in *The Route 66 Sticky Note Quest*: "Addy Settled on the Answer." (Your student will follow Addy and the Whitaker family as they travel into California, finding the final clue.)

Then follow the story up with some facts. You can choose which resource works best for your students:

- *Usborne Science Encyclopedia*: "Dynamics" pp. 122-123
- *Kingfisher Science Encyclopedia*: "Momentum" pp. 296-297
- *Usborne Illustrated Dictionary of Science:* "Dynamics" pp. 12-13
- *Everything You Need to Ace Science in One Big Fat Notebook:* "Motion" pp. 92-93

Discussion Questions

After all that reading, use the following questions to guide your discussion time with the students.

1. What is the study of dynamics? (Dynamics is the study of the relationship between the motion of an object and the forces that are acting on it.)
2. How is inertia measured? (Inertia is measured as mass.)
3. What is an impulse in physics? (An impulse in physics is a change in momentum or rather a force applied over time.)
4. What is the law of conservation of linear momentum? (The law of conservation of linear momentum says that when two or more objects exert forces on each other, the total momentum remains constant as long as no external forces affect them.)
5. How do rocket and jet engines move forward? (Both rocket and jet engines move forward due to a stream of high velocity gas exiting the rear of the engine.)

After the discussion time, have the students write a list of facts or a paragraph about what they read. Their writings should include at least a few facts from above.

NOTE-
Inertia and momentum are often confused or used interchangeably. However, the two are quite different. The amount of inertia force an object experiences is only based on the object's mass, whereas the force of momentum an object feels is dependent upon its mass and its speed. In other words, inertia is how much something resists changes in motion, while momentum increases or decreases with motion.

(Optional) Reading Selections

At another point in the week, the students can check out these optional books:

- *A Crash Course in Forces and Motion with Max Axiom, Super Scientist* by Emily Sohn and Steve Erwin

📖 *Motion at the Amusement Park (Amusement Park Science)* by Tammy Enz

Do: Experimenting with Science

Experiment: Investigating the Three Laws

Materials

You will need the following items:

- ✓ Jenga block with the eyehook from last week
- ✓ String
- ✓ 2 Toy cars
- ✓ Egg

Introduction

Read the following with your students, while they follow along in their lab journal.

Addy and the Whitaker girls reviewed the three laws of motion during a game of popcorn school. Isaac Newton built on Galileo's work on friction and motion through a number of experiments. These tests led to his development of the three laws of motion. The laws state:

- An object will not move, unless a force like a push or pull moves it. Once it is moving, an object will not stop moving unless it is forced to change.
- A force acting on an object will change its motion. The greater the force on an object, the greater the change in its motion.
- For every action, there is an equal but opposite reaction.

In today's experiment, you are going to do three tests where you will see each of the laws in action.

Directions

Have the students do the following experiment to investigate the three laws of motion under your supervision.

Motion Test #1

1. Take the Jenga block with the eyehook and tie the string to the block and place it on a smooth surface.
2. Pull the block along with a decent amount of force and then stop suddenly.
3. Observe what happens to the block.

Motion Test #2

1. Line up the two cars evenly on a flat surface.
2. Gently push your car forward, while your partner pushes his car forward with a greater force at the same time.
3. Observe what happens to the two cars.

Motion Test #3

1. Head outside with the egg.
2. Drop the egg onto the pavement from a height of four to five feet.
3. Observe what happens.

Results and Explanation

When they are done, read the results and explanation to the students.

> For test one, you should have seen that when they stopped suddenly, the block continued to move until the string went taut. Once the string was tight, the block bounced back a bit and eventually stopped. For test two, you should have seen that the car that was pushed with greater force moved faster and farther than the other car. For test three, you should have seen that the egg cracked and splattered on the pavement.

Discuss how the results from their experiment looked like or didn't look like the expected results. Then, read the following explanation to the students.

> In test one, we looked at inertia from the first law of motion. The block began to move because the student pulled the string attached to it. When you stopped suddenly, the block continued to move because the force of the sudden stop in the string has not acted on it yet. However, when then string went taut, the force is strong enough to change the motion of the block and eventually stop it. In test two, we saw that the force acting on the second car is greater than the force acting on the first one. This caused a greater change in the second car's motion, which means it goes faster and farther. In test three, the egg reached its terminal velocity before it hits the pavement. It was moving with such a force that the action of the sudden stop caused an equally violent reaction. This reaction has the force to break apart the egg and splatter it on the pavement.

(Optional) Hands-on Science Activities

At another point in the week, the students can do one or more of the following optional activities.

- **Egg Drop Carrier** – Have the students test how to slow down the inertia and momentum of a falling egg so that the "reaction" from the third law of motion doesn't result in a cracked egg. You will need a raw egg, various shock absorbing materials (such as cotton balls, newspaper, packing peanuts or fabric), a 1 quart plastic container (the type that fruit is typically packed in), and asking tape. Have the student begin by examining the different shock absorbing materials you have and using it to fill the 1 quart container in such a way that the egg will be protected as it falls. Be sure to have the egg on hand so that you can measure it and make sure the egg fits in the remaining space. Have the students tape across the top of the container to hold in the egg and the materials. Now, have them hold their containers over their heads and drop it. Observe what happens. Did the egg crack? (As the egg drops, its speed increases, which causes the force of momentum to increase. When it hits the ground, an equal shock force (Newton's 3rd Law of Motion) is sent back into the egg. If the egg was not protected, it would surely crack, like it did in the experiment from earlier this week. However, in this case, the materials surrounding the egg absorb most of the shock force, so that the force that the egg eventually felt was minimal.)

- **Second Law Worksheet** – Have the students practice calculating acceleration, mass, and force using the second law of motion using the worksheet in the Appendix on pp. 141-142 Answers:
    1. $a = 8 \text{ m/s}^2$
    2. F = 160,000 N
    3. m = 25 kg

Write: Keeping a Journal

Journaling Sheets

The students should have already recorded what they learned about the motion on their notes page (LJ p. 127) and completed the experiment sheet (LJ pp. 124-125). On the notes page, the students can also answer the bonus question. Here is an answer for your reference as you discuss this with your students.

- **Bonus Question** – Why does a delivery truck have more inertia than a bicycle? (A delivery truck has more mass than a bicycle, which leads to the truck having more inertia than the bicycle.)

Vocabulary

At some point in the week, have the students look up the following words and add the definitions to the unit vocabulary sheet on LJ pp. 94-97:

- **Inertia** – The tendency of an object to resist a change in its motion.
- **Mass** – The amount of matter in an object.
- **Momentum** – The tendency of an object to keep moving until a force stops it.
- **Weight** – The force with which an object's mass is pulled toward the center of the Earth.

Sketch

Here is the correctly labeled sketch. (LJ p. 126)

(Optional) Memory Work

At another point in the week, the students can work on the following memory work - Newton's Three Laws of Motion.

Three Laws of Motion

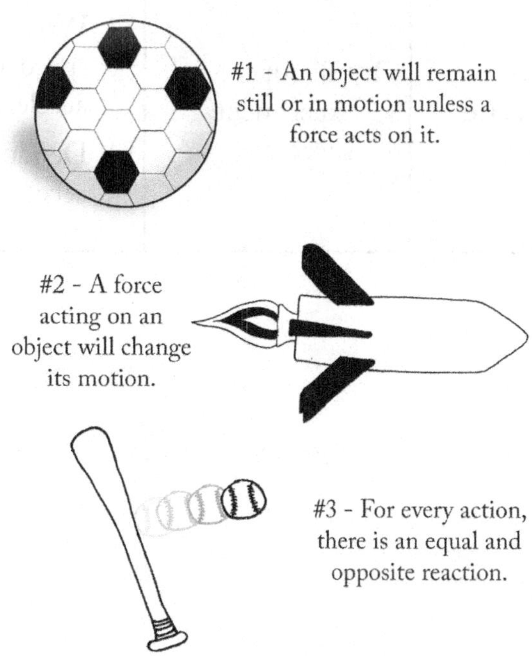

Lesson 17 Grid Schedule

Supplies Needed	
Experiments	• Cardboard or plastic track, Blocks or thick books, Toy car, Stopwatch
Optional Activities	• Experiment set-up, Sandpaper

Weekly Schedule

	Day 1	Day 2	Day 3	Day 4
Read	❑ Read Chapter 17: "Addy Hypothesized the Answer" in *The Route 66 Sticky Note Quest*	❑ Choose one of the encyclopedia readings to assign and discuss what was read using the questions provided	❑ *(Optional)* Choose another one of the encyclopedia readings to assign	❑ *(Optional)* Choose another one of the encyclopedia readings to assign
Do	❑ Do the "Will the height of the ramp affect a car's speed?" experiment		❑ *(Optional)* Complete the Acceleration Worksheet	❑ *(Optional)* Do the Friction and Acceleration activity
Write	❑ Fill out the experiment sheet	❑ Write about speed and acceleration ❑ Define acceleration	❑ Color and label the Acceleration Graph sketch ❑ Define speed and velocity	❑ *(Optional)* Work on memorizing Newton's Three Laws of Motion

Lesson 17 List Schedule

Essentials

Read

- O Read Chapter 17: "Addy Hypothesized the Answer" in *The Route 66 Sticky Note Quest.*
- O Choose one of the encyclopedia readings to assign and discuss what was read using the questions provided.

Do

- O Do the "Will the height of the ramp affect a car's speed?" experiment. Fill out the experiment sheet as you do the experiment.

Write

- O Write about speed and acceleration.
- O Look up acceleration, speed, and velocity. Add the definitions to the unit vocabulary sheet.
- O Label the Acceleration Graph sketch.

(Optional) Extras

Read

- O Choose another one (or two) of the encyclopedia readings to assign.

Do

- O Do the Friction and Acceleration activity.
- O Complete the Acceleration Worksheet.

Write

- O Work on memorizing Newton's Three Laws of Motion.

Supplies Needed	
Experiments	• Cardboard or plastic track, Blocks or thick books, Toy car, Stopwatch
Optional Activities	• Experiment set-up, Sandpaper

Lesson 17 Science Focus: Speed and Acceleration

Key Idea

Speed is the distance you cover over time.

Read: Gathering Information

Week's Readings

First, spark an interest by having the students read this week's story opener:

- Read Chapter 17 in *The Route 66 Sticky Note Quest*: "Addy Hypothesized the Answer." (Your student will follow Addy and the Whitaker family as they travel down the last few miles of Route 66.)

Then follow the story up with some facts. You can choose which resource works best for your students:

- *Usborne Science Encyclopedia*: "Motion" pp. 126-127
- *Kingfisher Science Encyclopedia*: No pages scheduled.
- *Usborne Illustrated Dictionary of Science:* "Motion" pp. 10-11
- *Everything You Need to Ace Science in One Big Fat Notebook*: "Speed, Velocity, and Acceleration" pp. 93-96

Discussion Questions

After all that reading, use the following questions to guide your discussion time with the students.

1. What is motion? (Motion is a change in position and orientation of an object.)
2. What is the difference between translational motion and rotational motion? (In translational motion, the object's center of mass moves from one place to another. In rotational motion, the movement is around the object's center of mass.)
3. What is the average speed? (Average speed is the distance traveled divided by the time taken.)
4. What is instantaneous speed? (Instantaneous speed is the speed at which an object is traveling at a given instant.)
5. How are speed and velocity related? (Velocity is the measurement of speed in a particular direction.)
6. What is acceleration? (Acceleration is a measure of how quickly velocity increases.)
7. What is deceleration? (Deceleration is a negative acceleration or a measure of how quickly velocity decreases.)

NOTE–
Often acceleration refers to an object that is speeding up, which is true to a point. In scientific terms, acceleration is related to velocity, not speed. Remember that velocity is a vector quantity, meaning it has both speed and direction. So, in physics, acceleration can be speeding up, slowing down, or just a change in direction.

After the discussion time, have the students write a list of facts or a paragraph about what they read. Their writings should include at least a few facts from above.

(Optional) Reading Selections

At another point in the week, the students can check out these optional books:

📖 *High-Speed Thrills: Acceleration and Velocity (The Science Behind Thrill Rides)* by Nathan Lepora

Do: Experimenting with Science

Experiment: Will the height of the ramp affect a car's speed?

Materials

You will need the following items:

- ✓ Cardboard or plastic track
- ✓ Blocks or thick books
- ✓ Toy car
- ✓ Stopwatch

Introduction

Read the following with your students, while they follow along in their lab journal.

> Olivia recites physics facts to help Addy get over motion sickness, but facts about speed don't do the trick. We know from Newton's laws of motion that a force must act on a vehicle to get it moving. We also know that the greater the force, the quicker the vehicle will go. In today's experiment, you are going to measure the time it takes for a toy car to go down a length of track. You will also vary the height of the track to see how different angles increase or decrease the force of gravity affecting the speed of the car.

Directions

Have the students do the following experiment to observe acceleration in action under your supervision.

1. Use the cardboard (or plastic track) to build a track that is 1 meter long.
2. Use the blocks (or books) to prop the track up so that one end of it is 15 centimeters higher than the other.
3. Now, use the stopwatch to measure the time it takes for the car to go from the top of the track to the end.
4. Record the time on the chart and then repeat two more times for a total of three trials at 15 centimeters.
5. Next, add several more blocks (or books) to prop the track up so that one end of it is 30 centimeters higher than the other.
6. Use the stopwatch to measure the time it takes for the car to go from the top of the track to the end.
7. Record the time on the chart and then repeat two more times for a total of three trials at 30 centimeters.

> NOTE- Make sure that the track is smooth, even, and free from debris that might slow down the progress of the car.

8. Finally, add a few more blocks (or books) to prop the track up so that one end of it is 45 centimeters higher than the other.
9. Use the stopwatch to measure the time it takes for the car to go from the top of the track to the end.
10. Record the time on the chart and then repeat two more times for a total of three trials at 45 centimeters.
11. Draw conclusions.

Results and Explanation

When they are done, read the results and explanation to the students.

> You should have seen that higher the track, the faster the average speed of the car.

Discuss how the results from their experiment looked like or didn't look like the expected results. Then, read the following explanation to the students.

> For the car to move down the track a force must either push or pull. In this experiment, the force of gravity overcomes the forces of friction and inertia to send the car heading down the track. As the height of the track increases, the force of gravity that pulls on the car also increases. This causes the speed of the toy car to increase as the height of the track increases.

(Optional) Hands-on Science Activities

At another point in the week, the students can do one or more of the following optional activities.

- **Friction and Acceleration (Experiment Repeat)** – Have the students cover the track with sandpaper to increase the friction. Then, have them repeat the experiment to see how the increased friction affects their results.
- **Acceleration Worksheet** – Have the students practice calculating acceleration using the worksheet in the Appendix on pp. 143-144.
  Answers
  1. 1.43 m/s2
  2. 5 m/s2
  3. 5.2 m/s2

Write: Keeping a Journal

Journaling Sheets

The students should have already recorded what they learned about speed and acceleration on their notes page (LJ p. 133) and completed the experiment sheet (LJ pp. 130-131). On the notes page, the students can also answer the bonus question. Here is an answer for your reference as you discuss this with your students.

- **Bonus Question** – What is one example of acceleration that doesn't involve a change in speed? (Driving at a constant speed around a bend in a road is one example.)

Vocabulary

At some point in the week, have the students look up the following words and add the definitions to the unit vocabulary sheet on LJ pp. 94-97:

- **Acceleration** – A change in an object's speed, direction, or both.
- **Speed** – The ratio of the distance an object moves to the amount of time the object moves.
- **Velocity** – The speed of an object in a particular direction.

Sketch

Here is the correctly labeled sketch. (LJ p. 132)

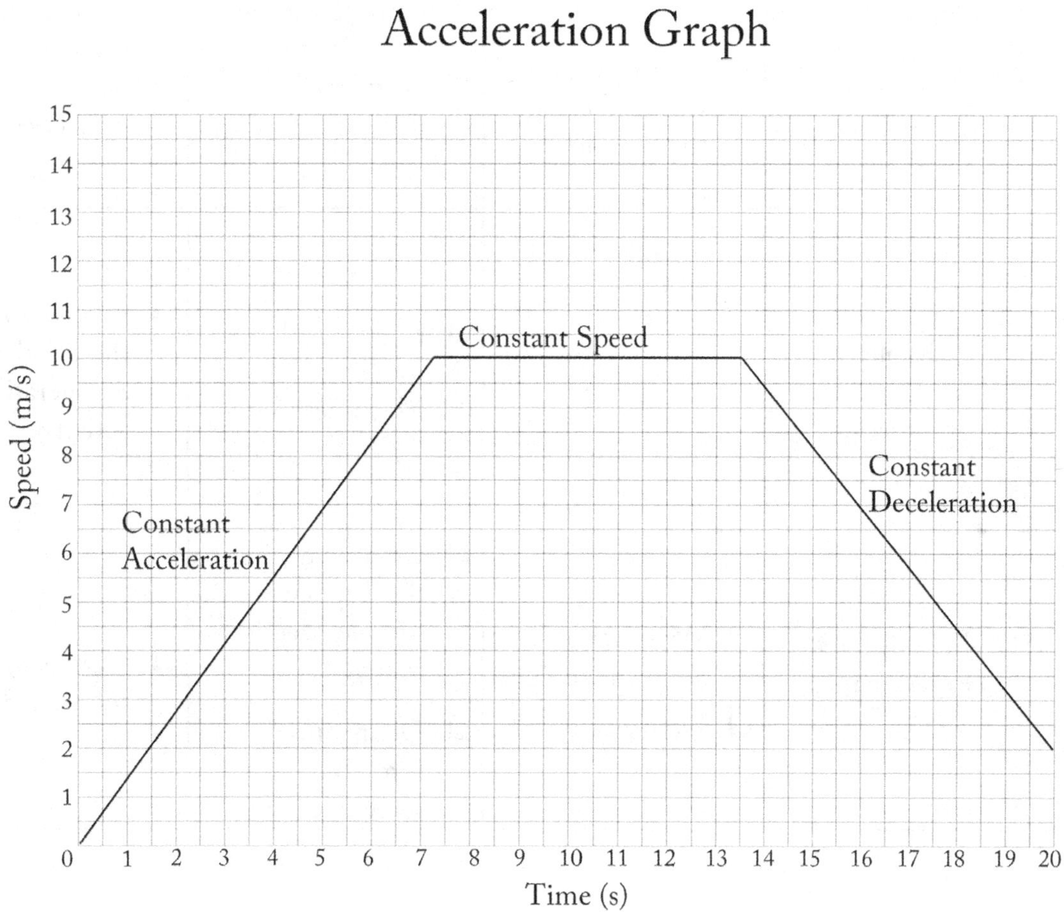

(Optional) Memory Work

At another point in the week, the students can work on the following memory work - Newton's Three Laws of Motion.

Lesson 18 Grid Schedule

Supplies Needed	
Experiments	• Materials will vary based on what you choose to build
Optional Activities	• Internet Access, Computer

Weekly Schedule				
	Day 1	**Day 2**	**Day 3**	**Day 4**
Read	❑ Read Chapter 18: "Addy Puzzled with Her Sisters" in *The Route 66 Sticky Note Quest*	❑ Choose one of the encyclopedia readings to assign and discuss what was read using the questions provided	❑ *(Optional)* Choose another one of the encyclopedia readings to assign	❑ Read "Epilogue: Addy Enjoyed the Day" in *The Route 66 Sticky Note Quest*
Do	❑ Do the "Build a Simple Machine" experiment		❑ *(Optional)* Play the Simple Machines Game	❑ *(Optional)* Do the Friction and Acceleration activity
Write	❑ Fill out the experiment sheet	❑ Write about simple machines ❑ Define fulcrum	❑ Color and label the Types of Simple Machines sketch ❑ Define input force and output force	❑ *(Optional)* Take the Unit 3 Test

Lesson 18 List Schedule

Essentials

Read

- O Read Chapter 18: "Addy Puzzled with Her Sisters" in *The Route 66 Sticky Note Quest.*
- O Choose one of the encyclopedia readings to assign and discuss what was read using the questions provided.
- O Read "Epilogue: Addy Enjoyed the Day" in *The Route 66 Sticky Note Quest.*

Do

- O Do the "Build a Simple Machine" experiment. Fill out the experiment sheet as you do the experiment.

Write

- O Write about simple machines.
- O Look up fulcrum, input force, and output force. Add the definitions to the unit vocabulary sheet.
- O Label the Types of Simple Machines sketch.

(Optional) Extras

Read

- O Choose another one (or two) of the encyclopedia readings to assign.

Do

- O Play the Simple Machines Game.

Write

- O Take the Unit 3 Test.

Supplies Needed	
Experiments	• Materials will vary based on what you choose to build
Optional Activities	• Internet Access, Computer

Lesson 18 Science Focus: Simple Machines

Read: Gathering Information

Key Idea

A simple machine is a tool to help you do work.

Week's Readings

First, spark an interest by having the students read this week's story opener:

- Read Chapter 18 in *The Route 66 Sticky Note Quest*: "Addy Puzzled with Her Sisters." (Your student will join Addy and the Whitaker family as they finish their journey down Route 66.)
- Read the epilogue of *The Route 66 Sticky Note Quest*: "Addy Enjoyed the Day."

Then follow the story up with some facts. You can choose which resource works best for your students:

- *Usborne Science Encyclopedia*: "Simple Machines" pp. 134-136
- *Kingfisher Science Encyclopedia*: "Ramps and Wedges" p. 300, "Levers and Pulleys" p. 301, "Wheels and Axles" p. 302
- *Usborne Illustrated Dictionary of Science:* "Machines" pp. 20-21
- *Everything You Need to Ace Science in One Big Fat Notebook*: "Simple Machines" pp. 121-126

Discussion Questions

After all that reading, use the following questions to guide your discussion time with the students.

1. What is a load? (A load is the force you need to overcome to move an object.) An effort? (An effort is the force needed to operate a machine.)
2. What can machines do? (Machines make physical work easier by reducing the effort needed or by magnifying the input force.)
3. What are simple machines? (Levers, wheels, pulleys, screws, gears, and inclined planes are all considered simple machines.) Give two examples of simple machines and describe how they work. (Answers will vary.)
4. What is a complex machine? (A complex machine is a combination of simple machines connected together to accomplish a job.) Give two examples of complex machines and describe how they work. (Answers will vary.)

After the discussion time, have the students write a list of facts or a paragraph about what they read. Their writings should include at least a few facts from above.

(Optional) Reading Selections

At another point in the week, the students can check out these optional books:

- *Simple Machines at the Amusement Park (Amusement Park Science)* by Tammy Enz
- *Simple Machines (Let's-Read-and-Find-Out Science 2)* by D. J. Ward

Do: Experimenting with Science

Experiment: Build a Simple Machine

Materials

You will need the following items:

- ✓ Materials will vary based on what you choose to build

Introduction

Read the following with your students, while they follow along in their lab journal.

Addy and her family talk about simple machines at a sculpture garden. A simple machine is a tool that you can use to help you do work. In other words, simple machines make the task of lifting or moving an object easier. There are seven main types of simple machines.

- Lever – A rigid bar that is free to move around at a fixed point. Levers are frequently used to lift things, i.e. using a flat piece of metal to pry open a paint can, using a wheelbarrow to move dirt, or using a shovel to life a large rock.
- Wheel and axle – Two disks or cylinders, each with a different radius. The wheel and axle simple machine is often used to turn or rotate things, i.e. using a screwdriver to turn a screw or using the steering wheel of a vehicle.
- Gears – Toothed wheels that interlock in pairs; each one helps to drive the next. Gears are often used to consistently drive an object, like a watch or clock.
- Inclined plane – A slanted surface that helps move objects up an incline. The inclined planes are typically ramps that can be used to move a heavy object from a lower level to a higher level.
- Wedge – A v-shaped object whose sides are two inclined planes. Wedges, like knives, axes, and zippers, make it easier to separate two objects.
- Screw – An inclined plane wrapped around a cylinder; this plane is also called the thread of the screw. These simple machines, including screws, nuts, and bolts, make it easier to drive an object into another object.
- Pulley – A rope that fits in the groove of a wheel. Pulleys make it easier to pull a heavy load directly upward.

In today's experiment, you are going to choose one of these types of simple machines to build.

Directions

Have the students do the following experiment to observe simple machines in action under your supervision.

1. Choose one of the types of simple machines (lever, wheel and axle, gear, inclined plane, wedge, screw, or pulley) to build.
2. Construct the machine and then test to see if it helps to

NOTE–
For ideas of simple machines the students can build easily at home, check out the following blog post:
- https://elementalscience.com/blogs/science-activities/3-ideas-for-building-simple-machines-at-home

perform work.

3. For instance, if you choose to build a lever, test it by using a stack of books. First, pick up the stack on your own, then use the lever to pick up the stack. Which was easier to do?
4. Draw conclusions.

(Optional) Hands-on Science Activities

At another point in the week, the students can do one or more of the following optional activities.

- **Simple Machines Game** – Have the students play the simple machines game from the Museum of Science and Industry in Chicago.

  https://edheads.org/game-page/simple-machines-interactive-educational-game/

Write: Keeping a Journal

Journaling Sheets

The students should have already recorded what they learned about simple machines on their notes page (LJ p. 139) and completed the experiment sheet (LJ pp. 136-137). There is no bonus question with this lesson.

Vocabulary

At some point in the week, have the students look up the following words and add the definitions to the unit vocabulary sheet on LJ pp. 94-97:

- **Fulcrum** – The point on which a lever rests or is supported and on which it pivots.
- **Input force** – The force you put into a machine.
- **Output force** – The force that a machine exerts on an object.

Sketch

See the correctly labeled sketch on the next page. (LJ p. 138)

(Optional) Forces and Motion Test

The students can complete the optional unit test found on pp. 160-164 in the Appendix. The answers for this test are found on pp. 130-131. A printable copy of the test can be downloaded for free from:

https://elementalscience.com/blogs/resources/sq200

Types of Simple Machines

Lever

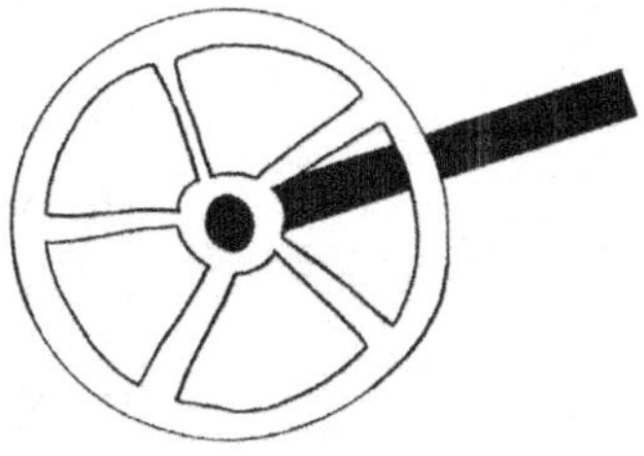

Wheel and Axle

Inclined Plane

Wedge

Screw

Pulley

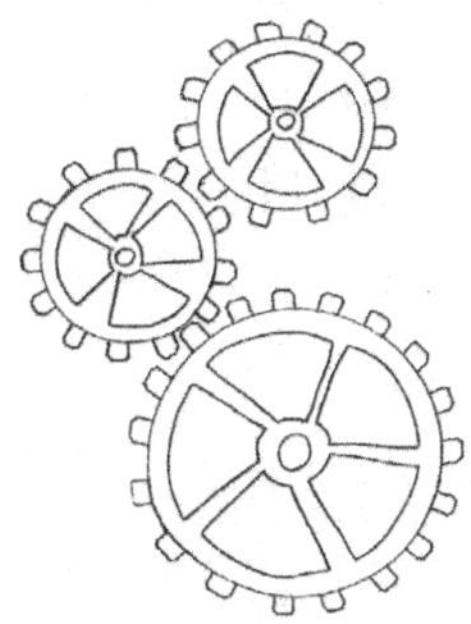

Gear

Forces and Motion Test Answers

Vocabulary Matching (2 points each)

1. J
2. F
3. B
4. S
5. X
6. P
7. E
8. K
9. M
10. T
11. H
12. I
13. R
14. W
15. L
16. N
17. A
18. V
19. U
20. O
21. C
22. G
23. D
24. O

Note -
If you need a digital copy of the unit test, you can download one for free from this program's resource page. Just head to:
https://elementalscience.com/blogs/resources/sq200

True and False (2 points each)

1. True
2. False (Gravitational force is the attraction between two objects that have mass. Magnetic force is the force between two moving charges.)
3. False (A streamlined design reduces drag so that the object can move more easily.)
4. True
5. True
6. False (An impulse in physics is a change in momentum or rather a force applied over time.)
7. False (Average speed is the distance traveled divided by the time taken.)
8. True
9. True
10. False (According to the law of conservation of energy, energy cannot be created or destroyed.)
11. True
12. True
13. True
14. False (A load is the force you need to overcome to move an object. An effort is the force needed to operate a machine.)

Short Answer

1. (2 points) A force influences the shape and/or motion of an object.

2. (4 points) The distance between the objects and the mass of the objects both affect the force of gravity.
3. (6 points) The first says: an object will stay still or keep on the same path at a constant speed unless a force pushes or pulls it. The second law says: the greater the force that acts on an object, the greater the change in movement. The third law says: for every action there is an equal and opposite reaction.
4. (2 points) Velocity is the measurement of speed in a particular direction.
5. (4 points) Kinetic energy is the energy of motion, but potential energy is the energy that an object has because of its position.
6. (6 points) If you spread force over a large area, you will reduce the amount of pressure exerted. If you concentrate the force over a small area, you will increase the amount of pressure exerted.

Total points: 100

SciQuest

Appendix Templates

Nomenclature Worksheet

Nomenclature, or naming compounds in chemistry, came about when scientists wanted a consistent way to refer to the compounds they were discovering. There are several basic things we need to know before we can learn how to name compounds in chemistry.

First, we use prefixes to distinguish compounds that have the same two elements in differing numbers. The first six prefixes, which are derived from Latin, are *mono* ("one"), *di* ("two"), *tri* ("three"), *tetra* ("four"), *penta* ("five"), and *hexa* ("six"). The prefix *mono*, is usually left out from the beginning of the first word of the name.

Second, many compounds have common names which we use instead. Over the years, it has become traditionally accepted to use these names over their systematic names. For example:

C_2H_6 - Ethane instead of dicarbon hexahydride

H_2O - Water instead of dihydrogen oxide

NH_3 - Ammonia instead of nitrogen trihydride

Finally, polyatomic ions, or ions with more than one atom, have special names. For example:

SO_4^{-2} - Sulfate

OH^- - Hydroxide

NO_3^- - Nitrate

CO_3^{-2} - Carbonate

PO_4^{-3} - Phosphate

When naming compounds, we use a different set of rules to name ionic and covalent compounds. Here are the rules to follow:

Ionic Compounds

(*metal and nonmetal*)

1. Write the name for the cation, which is the metal ion.
2. Write the name for the anion, which is the nonmetal ion or polyatomic ion. (***Note****—If the anion is a nonmetal atom, use the root name and add the suffix –ide.*)

Examples

NaCl – Sodium chloride

HCl - Hydrogen chloride

$CaSO_4$ – Calcium sulfate

Covalent Compounds

(*two nonmetals*)

1. Write the name of the first element. If the symbol is followed by a subscript of two or more, use the proper prefix to show the number.
2. Write the root name of the second element and add the suffix –ide. If the symbol is followed by a subscript of two or more, use the proper prefix to show the number.

Examples

CO - Carbon monoxide

N_2O_4 - Dinitrogen tetroxide

PCl_3 – Phosphorus trichloride

Ionic Compounds

MgO ______________________________

$NaBr$ ______________________________

Li_2S ______________________________

$MgSO_4$ ______________________________

$Be(OH)_2$ ______________________________

$Sr(NO_3)_2$ ______________________________

Covalent Compounds

CO_2 ______________________________

NO_2 ______________________________

SO_3 ______________________________

N_2S ______________________________

BF_3 ______________________________

P_2Br_4 ______________________________

Resultant Force Worksheet

Introduction

Objects have more than one force acting on them at any given time. If the forces are in the same direction, they add together. The effect of this addition on the object would be to accelerate, or move, it in that direction. If the forces are in opposing directions, they subtract or cancel each other out. The effect on the object depends upon the size of the opposing force. If the two forces are equal, they will balance each other out and the object will remain still. If one of the opposing forces is greater, the end result will be for the object to accelerate, or move, in that direction.

We can determine how an object will move by calculating the resultant force. The resultant force, which is also known as the net force, is the overall force acting on an object after all the forces are combined. To calculate the resultant force we use vector quantities to represent the forces. These vectors have both direction and size.

Sample Problems

Here are two sample problems for calculating the resultant force using vectors:

Resultant Force = 5N

The object will continue in the same direction.

Resultant Force = -4N

The object will begin moving in the opposite direction.

Problems

1.

Resultant Force = _______

The object will ____________________

____

2.

Resultant Force = _______

The object will ____________________

____

3.

Resultant Force = ______

The object will ____________________

____

4.

Resultant Force = ______

The object will ____________________

____

Work Worksheet

Introduction

Work is the transfer of energy that occurs when an object is moved or changed. In other words, work is the result of force applied over a distance. We use the following equation for calculating the amount of work in a given situation:

Work (W) = Force (F) • Distance (d)

The unit for work is the Joule (J), which stands for newton-meter (N•m). Now that we understand what work is and how to calculate it, let's look at a sample problem.

Sample Problems

A man is pushing a block with a 5 N force. He travels 15 meters. How much work did he do?

Force (F) = 5 N Distance (d) = 15 m Work (W) = ?

W = 5 • 15

W = 75 J

The man has done 75 J of work.

Problems

Now it is your turn to try a few problems!

1. A man used a 4 N force to push his stalled car the remaining 50 m to his house. How much work has the man done?

2. A person carries a bag of dog food that weighs 20 N up 12 m of stairs. How much work has that person done?

3. Once it has been filled, a hot air balloon weighs 900N. It quickly rises 700m. How much work did the hot air do?

4. A mechanical winch is attached to a bundle of lumber weighing 250 N. When you turn the winch on and use it to lift the lumber 4.2 m to the second floor window, how much work has the winch done?

Second Law Worksheet

Introduction

Newton's second law of motion states that:

> *A force acting on an object will change its motion. The greater the force on an object, the greater the change in its motion.*

This law can also be written as an equation which looks like:

Force (F) = mass (m) • acceleration (a)

We can use this equation to calculate the acceleration of an object when we know its mass and the force that has acted on it.

Sample Problems

Let's say you are pushing a grocery cart full of oranges that has a mass of 35 kg with a net force of 70 N. What would the acceleration of the cart be?

F = 70 N ($kg \cdot m/s^2$) m = 35 kg a = ?

Substitute and solve.

70 N = (35 kg) a

$$a = \frac{70\ N\ (\text{or}\ kg \cdot m/s^2)}{35\ kg}$$

$a = 2\ m/s^2$

Problems

Now it is your turn to try a few problems!

1. A girl kicks a 2.5 kg soccer ball with a net force of 20 N. What would the acceleration of the ball be?

2. A truck pulls a trailer with mass of 4000 kg up a hill at a rate of 40 m/s^2. What is the net force that acted on the trailer?

3. A 12.5N force accelerates a boy in a toy car at 0.5 m/s^2. What is the mass of the boy?

ACCELERATION WORKSHEET

INTRODUCTION

Acceleration is the change in velocity over a period of time. We can calculate the acceleration of an object using the following formula:

$$\text{Acceleration (a)} = \frac{\text{change in velocity } (v_f - v_i)}{\text{total time (t)}}$$

If we have a positive acceleration, it means that the object is speeding up. If we have a negative acceleration, the object is slowing down.

SAMPLE PROBLEMS

So, let's say we have a toy car that is stopped at the top of a ramp, which means its velocity is 0 m/s. The car is pushed down the ramp. After 3 seconds, the velocity of the toy car is 30 m/s. What is the car's acceleration?

v_f = 30 m/s v_i = 0 m/s t = 3 s a = ?

Substitute and solve.

$$a = \frac{(30 \text{ m/s} - 0 \text{ m/s})}{3 \text{ s}}$$

$$a = \frac{30 \text{ m/s}}{3 \text{ s}}$$

$$a = 10 \text{ m/s}^2$$

PROBLEMS

Now it is your turn to try a few problems!

1. A student is sitting quietly enjoying the park, when suddenly he is scared by a spiny tailed lizard! He jumps off the park bench on which he is sitting and takes off running. He reaches the gas station down the road in 7 seconds. At that point he is moving 10 m/s. What is the student's acceleration?

2. A vehicle is driving along the road at 25 m/s. It hits a patch of ice and its speed increases to 35 m/s in 2 seconds. What is the vehicle's acceleration?

3. A female cheetah is chasing a gazelle at a speed of 28 m/s. She misses and ends the chase. Over the next 5 seconds, she slows down to a restful pace of 2 m/s. What is the cheetah's acceleration?

SciQUEST

GLOSSARY

Level 200 Glossary

A

- **Acceleration** – A change in an object's speed, direction, or both.
- **Air Resistance** – The force that air exerts on an object as it falls.
- **Anode** – The positively charged electrode by which current enters the cell.

B

- **Balance** – A state of equilibrium when the forces acting on an object cancel each other out; also known as a zero resultant force.
- **Boiling Point** – The temperature at which a substance turns from a liquid into a gas.

C

- **Cathode** – The negatively charged electrode by which current leaves the cell.
- **Chemical Reaction** – An interaction between the atoms of two substances where the atoms rearrange to form two new molecules.
- **Chromatography** – A method of separating the substances in a mixture by the rate they move through or along a medium, such as filter paper.
- **Cleavage** – The splitting of a crystal along a certain plane.
- **Conductivity** – The measure of a substance's ability to conduct heat or electricity.
- **Covalent Bonding** – A chemical bond between two atoms, in which each atom shares an electron.

D

- **Density** – A measure of the amount of matter in a substance compared to its volume.
- **Diffusion** – The spreading out of a gas to fill the available space of the container it is in.

E

- **Elasticity** – The ability of a substance to stretch and then return to its original shape.
- **Electrolyte** – A substance that conducts electricity when it is in solution.
- **Endothermic Reaction** – A chemical reaction that takes in heat.
- **Energy** – The ability to do work.
- **Energy Conversion** – The process of changing one form of energy into another.
- **Exothermic Reaction** – A chemical reaction that gives off heat.

F

- **Fluid** – A substance that assumes the shape of the container it is in, such as a liquid or gas.
- **Force** – A push or pull that acts on an object.
- **Force Field** – The area in which a force can be felt.
- **Freezing Point** – The temperature at which a substance turns from a liquid into a solid.
- **Friction** – A force that opposes the motion of objects that touch as they move past each other.
- **Fulcrum** – The point on which a lever rests or is supported and on which it pivots.

G

- **Gravity** – The force that acts between two masses; it is an attractive force.

H

I

- **Immiscible** – Liquids that cannot be blended together.
- **Inertia** – The tendency of an object to resist a change in its motion.
- **Input Force** – The force you put into a machine.
- **Intermolecular Forces** – The forces that exist between molecules, which explain properties like elasticity, surface tension, and viscosity.
- **Ion** – An atom or group of atoms that has become charged by gaining or losing one or more electrons.
- **Ionic Bonding** – A strong chemical bond that is formed by the attraction between two ions of opposite charges.

J

K

- **Kinetic Energy** – The energy of an object in motion; it depends upon the object's mass and speed.
- **Kinetic Theory** – The theory that states that as the temperature rises, particles move faster and therefore take up more space.

L

- **Lattices** – The regular arrangement of repeating patterns of atoms, molecules, or ions in a solid structure.

M

- **Mass** – The amount of matter in an object.
- **Metallic Bonding** – A chemical bond where positive metal ions form a lattice structure with freely moving electrons between them.
- **Miscible** – Liquids that can be blended together.
- **Mixture** – A combination of two or more elements or compounds that are not chemically combined.
- **Molecule** – A substance that is formed when two or more atoms chemically join together.
- **Momentum** – The tendency of an object to keep moving until a force stops it.

N

- **Newton** – The measurement of force; one newton is the force is takes to move a one kilogram object at one meter per second squared ($1 \text{ N} = 1 \text{ kg} \cdot 1 \text{ m/s}^2$).

O

- **Output Force** – The force that a machine exerts on an object.

P

- **Plasma** – The fourth state of matter that exists only at very high temperatures.
- **Potential energy** – The energy that an object has stored; it depends upon the object's weight and height.
- **Pressure** – The amount of force pushing on a given area.
- **Product** – A new substance produced in a chemical reaction.

Q

R

- **Reactant** – The substance present at the beginning of a chemical reaction.

S

- **Solubility** – The ability of a solute to be dissolved.
- **Solute** – The substance that dissolves in the solvent to form a solution.
- **Solution** – A homogenous mixture of two or more substances.
- **Solvent** – The substance in which the solute dissolves to form a solution, typically a liquid.

- **Speed** – The ratio of the distance an object moves to the amount of time the object moves.
- **Sublimation** – Occurs when a substance changes directly from a solid to a gas without changing into a liquid.

T

- **Temperature** – A measure of the heat energy that a substance contains.
- **Terminal Velocity** – The point at which the force acting on an object of air resistance is equal to the force of gravity acting on the object.

U

V

- **Velocity** – The speed of an object in a particular direction.
- **Volume** – A measurement of the space occupied by a substance.

W

- **Weight** – The force with which an object's mass is pulled toward the center of the Earth.
- **Work** – The transfer of energy that occurs when a force moves or changes an object.

X

Y

Z

SciQuest

Unit Tests

Matter Unit Test

Vocabulary Matching

1.___ Boiling Point

2.___ Cleavage

3.___ Conductivity

4.___ Density

5.___ Diffusion

6.___ Elasticity

7.___ Freezing Point

8.___ Intermolecular forces

9.___ Kinetic Theory

10. ___ Lattices

11. ___ Plasma

12. ___ Pressure

13. ___ Sublimation

14. ___ Temperature

15. ___ Volume

A. A measure of the amount of matter in a substance compared to its volume.

B. The temperature at which a substance turns from a liquid into a solid.

C. The ability of a substance to stretch and then return to its original shape.

D. The temperature at which a substance turns from a liquid into a gas.

E. The fourth state of matter that exists only at very high temperatures.

F. The measure of a substance's ability to conduct heat or electricity.

G. The amount of force that pushes on a given area.

H. The theory that states that as the temperature rises, particles move faster and therefore take up more space.

I. Occurs when a substance changes directly from a solid to a gas without changing into a liquid.

J. A measure of the heat energy that a substance contains.

K. The regular arrangement of repeating patterns of atoms, molecules, or ions in a solid structure.

L. A measure of the space occupied by a substance.

M. The spreading out of a gas to fill the available space of the container it is in.

N. The splitting of a crystal along a certain plane.

O. The forces that exist between molecules, which explain properties like elasticity, surface tension, and viscosity.

True and False

1. ______________________ Substances can change physical state when heated or cooled, or when the energy of the particles is increased or decreased.

2.____________________ The melting point is the point at which a liquid changes into a gas and the boiling point is the point at which a solid changes into a liquid.

3.____________________ Two methods of crystallization are allowing a solvent to evaporate and placing a seed crystal in a supersaturated solution.

4.____________________ The four main types of lattice structures are giant atomic, giant ionic, giant metallic, and molecular lattices.

5.____________________ Cohesion is the intermolecular force of attraction between molecules within different substances, while adhesion is the intermolecular force of attraction between molecules within the same substance.

6.____________________ Avogadro's law states that equal volumes of gases at the same temperature and pressure contain the same number of molecules.

7.____________________ An ideal gas behaves in an ideal way.

8.____________________ Stp stands for super-trans-phosphorus.

Short Answer

1. What are the three states of matter typically found on Earth? (Be sure to give a characteristic of each.)

2. What causes crystals to form a definite shape?

3. What is surface tension?

4. What is Boyle's Gas Law?

5. What is Charles's Gas Law?

Solutions and Reactions Unit Test

Vocabulary Matching

1. ___ Anode
2. ___ Cathode
3. ___ Chemical Reaction
4. ___ Chromatography
5. ___ Covalent Bonding
6. ___ Electrolyte
7. ___ Endothermic Reaction
8. ___ Exothermic Reaction
9. ___ Immiscible
10. ___ Ion
11. ___ Ionic Bonding
12. ___ Metallic Bonding
13. ___ Miscible
14. ___ Mixture
15. ___ Molecule
16. ___ Product

A. Liquids that can be blended together.

B. The substance present at the beginning of a chemical reaction.

C. A homogenous mixture of two or more substances.

D. A combination of two or more elements or compounds that are not chemically combined.

E. A strong chemical bond that is formed by the attraction between two ions of opposite charges.

F. An atom or group of atoms that has become charged by gaining or losing one or more electrons.

G. A substance that is formed when two or more atoms chemically join together.

H. A substance that conducts electricity when it is in solution.

I. A chemical bond where positive metal ions form a lattice structure with freely moving electrons between them.

J. The positively charged electrode by which current enters the cell.

K. An interaction between the atoms of two substances where the atoms rearrange to form two new molecules.

L. Liquids that cannot be blended together.

M. The ability of a solute to be dissolved.

N. A chemical reaction that gives off heat.

O. The substance that dissolves in the solvent to form a solution.

P. A chemical reaction that takes in heat.

17. ___ Reactant

18. ___ Solubility

19. ___ Solute

20. ___ Solution

21. ___ Solvent

Q. A chemical bond between two atoms, in which each atom shares an electron.

R. A method of separating the substances in a mixture by the rate they move through or along a medium, such as filter paper.

S. The substance in which the solute dissolves to form a solution, typically a liquid.

T. The negatively charged electrode by which current leaves the cell.

U. A new substance produced in a chemical reaction.

True or False

1.____________________ A mixture is composed of elements or compounds that have been chemically joined.

2. ____________________ Chemists say that a substance is pure when it contains only one element or compound.

3. ____________________ Water is a good solvent because it is a polar molecule.

4.____________________ The solubility of a solid increases as the temperature increases. The solubility of a gas decreases as the temperature increases.

5. ____________________ Evaporation is the easiest method for separating a suspension.

6. ____________________ Distillation is a method of separating a mixture that it best used when you need to separate the solvent from the solution.

7. ____________________ An electrode is a piece of metal or graphite that allows the electrical current to flow into the electrolyte.

8. ____________________ Electrorefining is the process of using electrolysis to purify a metal, such as copper.

9.____________________ A cation has a positive charge. An anion has a negative charge.

10. ____________________ Chemical bonding always involves the innermost electrons of an atom.

11. ______________________ The substances present at the beginning of the reaction are the products. The substances present at the end of the reaction are the reactants.

12. ______________________ In a chemical reaction, bonds are broken and made, which requires energy.

Short Answer

1. What is the law of constant composition?

2. What is the difference between concentrated and dilute solutions?

3. Name two types of separating a mixture and why you would choose each of the methods.

4. What is an electrolyte, and why do these substances conduct electricity?

5. Briefly explain ionic bonding, covalent bonding, and metallic bonding.

6. What is the difference between exothermic and endothermic reactions?

Forces and Motion Unit Test

Vocabulary Matching

1.____ Acceleration

2.____ Air resistance

3.____ Balance

4.____ Energy

5.____ Energy conversion

6.____ Fluid

7.____ Force

8.____ Force field

9.____ Friction

10. ____ Fulcrum

11. ____ Gravity

12. ____ Inertia

13. ____ Input force

14. ____ Kinetic energy

15. ____ Mass

16. ____ Momentum

17. ____ Newton

18. ____ Output force

19. ____ Potential energy

A. The measurement of force; 1 Newton (N) is the force is takes to move a one kilogram object at 1 meter per second squared ($1 \text{ N} = 1 \text{ kg} \cdot 1 \text{ m/s}^2$).

B. A state of equilibrium when the forces acting on an object cancel each other out; also known as a zero resultant force.

C. The point at which the force acting on an object of air resistance is equal to the force of gravity acting on the object.

D. The force with which an object's mass is pulled toward the center of the Earth.

E. A push or pull that acts on an object.

F. The force that air exerts on an object as it falls.

G. The speed of an object in a particular direction.

H. The force that acts between two masses; it is an attractive force.

I. The tendency of an object to resist a change in its motion.

J. A change in an objects 'speed, direction, or both.

K. The area in which a force can be felt.

L. The amount of matter in an object.

M. A force that opposes the motion of objects that touch as they move past each other.

N. The tendency of an object to keep moving until a force stops it.

O. The ratio of the distance an object moves to the amount of time the object moves.

P. A substance that assumes the shape of the container it is in, such as a liquid or gas.

Q. The transfer of energy that occurs when a force moves or changes an object.

R. The force you put into a machine.

S. The ability to do work.

20. ____ Speed	T. The point on which a lever rests or is supported and on which it pivots.
21. ____ Terminal velocity	U. The energy that an object has stored; it depends upon the object's weight and height.
22. ____ Velocity	V. The force that a machine exerts on an object.
23. ____ Weight	W. The energy of an object in motion; it depends upon the object's mass and speed.
24. ____ Work	X. The process of changing one form of energy into another.

True or False

1. ____________________ A vector quantity has both magnitude and direction.

2. ____________________ Electric force is the attraction between two objects that have mass. Nuclear force is the force between two moving charges.

3. ____________________ A streamlined design increases drag so that the object can move more easily.

4. ____________________ Weight is not constant.

5. ____________________ Inertia is measured as mass.

6. ____________________ An impulse in physics is an instant force applied only once.

7. ____________________ Average speed is the speed at which an object is traveling at a given instant.

8. ____________________ Acceleration is a measure of how quickly velocity increases or decreases.

9. ____________________ Energy is the ability to do work, and work cannot be done without energy.

10. ____________________ Energy can be created and destroyed at will.

11. ____________________ Upthrust is the upward force an object in a fluid experiences, and it is equal to the weight of the fluid displaced.

12. ____________________ An object will float if the upthrust is equal to the object's weight.

13. ____________________ A complex machine is a combination of simple machines connected together to accomplish a job.

14. ____________________ An effort is the force you need to overcome to move an object. A load is the force needed to operate a machine.

Short Answer

1. What does a force do?

2. What two things affect the force of gravity between two objects?

3. What are Newton's three laws of motion?

4. How are speed and velocity related?

5. What is the difference between kinetic and potential energy?

6. How are pressure, force, and area related?

Made in the USA
Monee, IL
16 September 2026

63296739R00092